Selenium and Human Health

Edited by Volkan Gelen,
Adem Kara and Abdulsamed Kükürt

Published in London, United Kingdom

Selenium and Human Health
http://dx.doi.org/10.5772/intechopen.105312
Edited by Volkan Gelen, Adem Kara and Abdulsamed Kükürt

Contributors
Muhammad Imran, Zhikun Chen, Shah Rukh, Ayaz Mehmood, Wang Weixi, Waleed Asghar, Farhan Iftikhar, Mayowa J. Adeniyi, Cynthia Okeke, Akhabue K Okojie, Ayoola Awosika, Irina Alexandrovna Savvina, Anna Olegovna Petrova, Hasaybat Salimbekovna Nucalova, Kristina M. Bykova, Irina Varlamovna Tkebuchava, Soledad García-Morales, Janet María León-Morales, Víctor García-Gaytán, Luis Guillermo Terreros-Rosales, Shukurlu Yusif Hajibal, Huseynov Tokay Maharram, Mehmet Başeğmez, Muhammed Fatih Doğan, Kumar Venkitanarayanan, Poonam Gopika Vinayamohan, Divya Jopseph, Leya Susan Viju, Volkan Gelen, Adem Kara, Abdulsamed Kükürt

First published in London, United Kingdom, 2023 by IntechOpen
IntechOpen is the global imprint of INTECHOPEN LIMITED, registered in England and Wales, registration number: 11086078, 5 Princes Gate Court, London, SW7 2QJ, United Kingdom

British Library Cataloguing-in-Publication Data
A catalogue record for this book is available from the British Library

Additional hard and PDF copies can be obtained from orders@intechopen.com

Selenium and Human Health
Edited by Volkan Gelen, Adem Kara and Abdulsamed Kükürt
p. cm.
Print ISBN 978-1-83768-910-1
Online ISBN 978-1-83768-911-8
eBook (PDF) ISBN 978-1-83768-912-5

Meet the editors

Dr. Volkan Gelen is a physiology specialist who received his veterinary degree from Kafkas University, Turkey, in 2011. From 2011 to 2015, he worked as an assistant in the Department of Physiology, Faculty of Veterinary Medicine, Atatürk University, Turkey. In 2016, he joined Kafkas University as an assistant professor where he has been engaged in various academic activities. Dr. Gelen has sixty journal articles and twenty poster presentations to his credit. His research interests include physiology, the endocrine system, cancer, diabetes, cardiovascular diseases, and isolated organ bath system studies.

Dr. Adem Kara completed his doctorate in the Department of Histology and Embryology at the Faculty of Medicine in 2012 at Atatürk University. He worked as a lecturer in the Faculty of Veterinary Medicine, Department of Histology and Embryology, Atatürk University, Turkey. Dr. Kara has authored and co-authored fifty-seven journal papers, four book chapters, and thirty other types of publications.

Dr. Abdulsamed Kükürt graduated from Uludağ University, Turkey. In 2019, he obtained a Ph.D. in Biochemistry from the Institute of Health Sciences, Kafkas University, Turkey, where he is currently an assistant professor in the Department of Biochemistry. He has published twenty-eight research articles in academic journals, eleven book chapters, and thirty-seven papers. Dr. Kükürt has participated in ten academic projects and is a reviewer for many academic journals.

Contents

Preface

Selenium is an antioxidant substance that contributes to the growth and development of cells and strengthens the immune system. Selenium is found in the structure of selenoproteins, which help prevent cellular damage caused by free radicals.

It is known that trace elements have a role in various diseases, especially in wound healing and correction of the immune response. It has been determined by various studies that low selenium levels may be associated with mortality and morbidity.

In most diseases, the organism is under oxidative stress. Selenium-containing enzymes protect cells against lipid peroxidation and play a role in regulating inflammatory events. Therefore, administration of selenium may prevent cellular damage in patients.

This book contains comprehensive information about the effects of selenium and its various components, which are extremely important for human health, in the organism, its role in various diseases, and especially its importance in human health. The book includes contributions from leading scientists and experts in the field. In addition, the book provides a vital foundation and opportunities for future research on the relationship between selenium and human health. We would like to express our gratitude to all the authors for their excellent contributions.

Volkan Gelen,
Faculty of Veterinary Medicine,
Department of Physiology,
Kafkas University,
Kars, Turkey

Adem Kara,
Faculty of Science,
Department of Molecular Biology and Genetics,
Erzurum Technical University,
Erzurum, Turkey

Abdulsamed Kükürt,
Deparment of Biochemistry,
Kafkas University,
Kars, Turkey

Section 1

The Effects of Selenium in the Organism

Chapter 1

Synthesis and Types of Selenoproteins and Their Role in Regulating Inflammation and ER Stress Signaling Pathways: Overview

Volkan Gelen, Adem Kara and Abdulsamed Kükürt

Abstract

Selenium (Se) is one of the trace elements that play an important role in many biological processes in the living body. Selenium acts in the body mainly in its forms called selenoprotein. Selenoproteins play a role in various events such as oxidative stress, immunity, cancer, inflammation, and endoplasmic reticulum stress. In selenium deficiency, the expression of selenoproteins and thus their activity decrease. In this case, some reactions such as increased oxidative stress, weakened immunity, endoplasmic reticulum stress, and inflammation cannot be prevented. The main source of selenium is food, and a diet poor in selenium causes selenium and therefore selenoprotein deficiency. This chapter will present information about the synthesis of selenoproteins and their role, especially in inflammation and endoplasmic reticulum stress response.

Keywords: selenium, selenoproteins, ER stress, inflammation, oxidative stress

1. Introduction

Selenium (Se) is a trace element and must be taken from outside. Selenium was first discovered in 1817 [1]. Research on the effects of Se on the organism has gained momentum over time. Se has an important role in the regulation of many functions in the organism such as reproductive physiology, muscle functions, cardiovascular system, nervous system, and immune system [2]. Selenium is mainly found in many products such as soil, water, vegetables, fruits, meat, milk, eggs, and fish [3, 4]. Both excess and deficiency of selenium cause some problems [2]. Selenium deficiency causes a number of problems such as acute heart failure, arrhythmia, muscular dystrophy, short stature, and short extremities [5–7]. On the other hand, excessive intake of Se causes hair loss, deterioration in nail structure, and nervous system anomalies [1, 2]. In other words, as it can be understood, excess and deficiency of selenium cause a number of problems. Selenium can be taken into the body in organic Se and inorganic forms. The inorganic forms of selenium are mostly selenate and selenite. Its organic form is selenomethionine (Se-Met) and selenocysteine (Sn) [8, 9]. Sec and Se-Met have many biological roles. The structures formed by proteins that combine with Se are called selenoproteins [10]. Selenoproteins are also involved in various biological functions such as maintaining homeostasis in the organism, oxidative stress, hormone release, regulation of the immune system, inflammation, and stress on the endoplasmic reticulum [11]. Selenium or selenoprotein deficiency is generally due to insufficient intake of foods [12]. The most common forms of Se are selenate, selenite, Sec, and Se-Met [13]. These forms are very active in homeostasis. In addition, it has been stated that they have many effects on cancer [14]. In line with this information, in this section, we aimed to explain the mechanism of action by discussing the synthesis of selenoprotein forms of Se, which is of such importance for the organism, their types, and their roles in inflammation and ER-stress.

2. Synthesis of selenoproteins

Selenium shows its effect on living things through selenoproteins. Its main biological form is selenocysteine, and its synthesis begins with the binding of the serine amino acid to tRNA [15]. Selenocysteine is similar to cysteine, but it has a selenium atom instead of sulfur in its structure and is ionized at physiological pH. In the study, replacing selenocysteine with cysteine dramatically reduces enzyme activity [16–19]. This supports the critical role of the ionized selenium atom [20]. Selenoproteins contain one or more selenocysteine residues in their primary structure [21]. According to current information, all selenoproteins, except Selenoprotein P, take part in redox reactions, are located in the catalytic regions of enzymes, and show enzymatic activity. Although selenoproteins have many similar functions in general, their amino acid sequences, tissue distributions of enzymatic activities, and interactions with other molecules vary widely [18, 19], looking at the selenoprotein synthesis steps (**Figure 1**).

3. Types of selenoproteins

Selenium can enter the body in various forms, but its absorption is mainly in the form of selenoprotein [3]. As a result of various studies, 25 selenoproteins, 5 of which are glutathione, have been isolated in humans. These selenoproteins are selenium phosphorylate synthetase (SPS), selenoprotein S (SELENOS), selenoprotein H (SELENOH), peroxidases (GPXs), 3 thioredoxin reductases (TrxRs), 3 iodothyronine deiodinases (DIOs), selenoprotein P (SELENOP), selenoprotein W (SELENOW), selenoprotein M (SELENOM), SELENON), selenoprotein I (SELENOI), selenoprotein K (SELENOC), selenoprotein N (selenoprotein O (SELENOO), selenoprotein T (SELENOT), selenoprotein 15 (15 kDa), selenoprotein R (SELENOR), and selenoprotein V (SELENOV) [17, 21]. Selenoproteins are found in various parts of the cell such as mitochondria, endoplasmic reticulum, nucleus, cell membrane, and Golgi membrane. And where they are found, they have various functions such as antioxidant, anti-inflammatory, hormone metabolism, and regulation of ER stress [22, 23]. The types, names, locations, and functions of some human selenoproteins are summarized in **Table 1**.

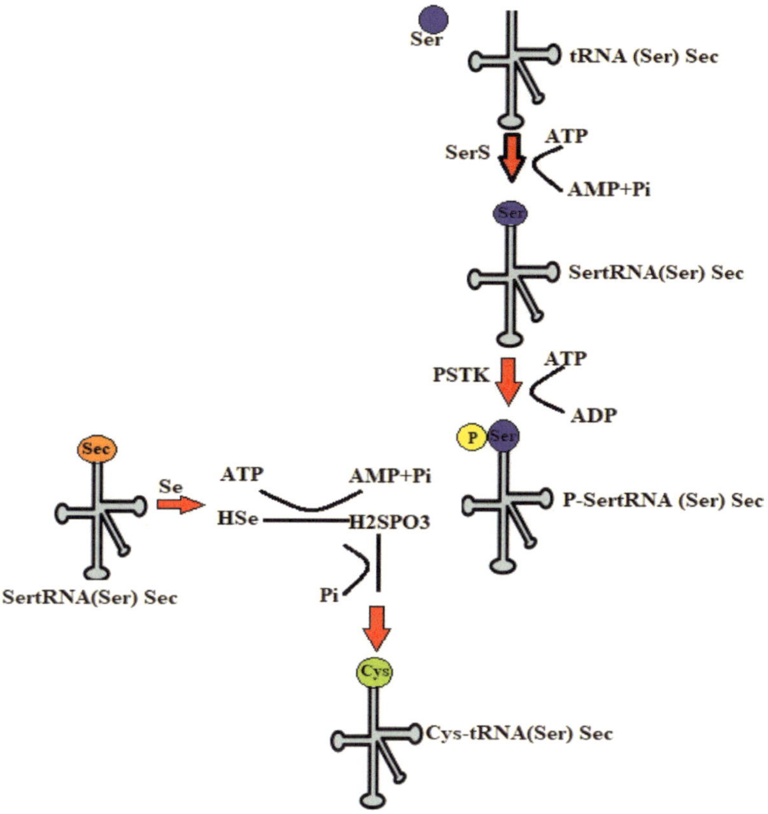

Figure 1.
Synthesis of selenoproteins [20].

4. Roles of selenoproteins in inflammation

Glutathione peroxidase, which protects cells against oxidative damage, is found in the cytoplasm of cells and originates from hydrogen peroxide (H_2O_2). In this way, it prevents the formation of OH from H2O2. Glutathione peroxidase has four protein subunits. Each of the subunits contains a selenium atom. Two main types of glutathione peroxidase enzymes have been identified. The first is selenium-dependent glutathione peroxidase (Se-GPx), which has selenium in its active site. Selenium-dependent glutathione peroxidase has an active role against organic hyper oxides and H_2O_2. Selenium-independent glutathione peroxidase (GST) is known to be more active in the formation of organic hydroperoxides. GPX1 suppresses inflammation in the cell by affecting proinflammatory cytokines and preventing ROS accumulation. Here, the Nrf2/ARE pathway plays an important role [24]. GPX also catalyzes glutathione in various tissues, preventing peroxidation of free radicals and preventing oxidative stress-induced DNA damage in the cell [17, 25–28]. Some studies have shown that Se supplementation increases GPx and SOD activity and decreases MDA levels [29]. In these studies, it inhibits cell inflammation and apoptosis by suppressing ROS-mediated NF-κB production [24, 30]. It has been determined that GPx2 and GPx1 suppress inflammation in intestinal epithelial cells [31–33]. It has been found that vascular inflammation is stimulated in Se deficiency [20]. In another study, it was shown

Selenoproteins	Short name	Main function	Location
Glutathione peroxidase 1	GPx1/cGPx	Antioxidant (detoxification of hydrogen peroxide)	Cytoplasm
Glutathione peroxidase 2	GPx2/GI (gastrointestinal)-GPx2	Antioxidant (detoxification of hydrogen peroxide)	Cytoplasm
Glutathione peroxidase 3	GPx3	Antioxidant (detoxification of hydrogen peroxide)	Secreted
Glutathione peroxidase 4	GPx4/PH (Phospholipid hydroperoxide)-GPx	Antioxidant protects against lipid peroxidation	Cytoplasm, mitochondria, nucleus, and memberanes
Glutathione peroxidase 6	6 GPx6	Antioxidant (detoxification of hydrogen peroxide)	Secreted
Thioredoxin reductase 1	TR1	Reduction of thioredoxins and other substrates	Cytoplasm, nuclear
Thioredoxin reductase 2	TR2	Reduction thioredoxin disulfide bond isomerization, thioredoxin/glutaredoxin/glutathione reductase	Mitochondria
Thioredoxin reductase 3	TR3		Mitochondrial, nuclear, cytoplasm?
Deiodinase type I	Dio1	Thyroid hormone metabolism	ER membrane
Deiodinase type II	Dio2	Thyroid hormone metabolism	Membrane?
Deiodinase type III	Dio3	Thyroid hormone catabolism	Cell and endosome membrane
Selenophosphate	SPS2	Conversion of selenide to selenophosphate	Unknown
Selenoprotein P	SePP	Se transport and delivery/anti-oxidant	Secreted
Selenoprotein W	SelW antioxidant?	Antioxidant?	Cytoplasm
Selenoprotein K	SelK ER	Antioxidant? regulates Ca2þ flux	ER
15 kDa Selenoprotein	SeP15	Protein folding	ER
Selenoprotein S	SePS/SelS	Inflammatory response, regulation cytokine production, protection against ER-stress-induced apoptosis	ER
Selenoprotein M	SelM	Antioxidant? Or calcium homeostasis?	ER
Selenoprotein N	SelN	Antioxidant? calcium homeostasis? role in muscle function	ER membrane
Selenoprotein T	SelT	Unknown	Golgi/ER
Selenoprotein H	SelH	Nucleolar oxidoreductase, nuclear-localized DNA-binding protein?	Nuclear, nucleolar?
Selenoprotein I	SelI	Unknown	Unknown

Selenoproteins	Short name	Main function	Location
Selenoprotein O	SelO	Unknown	Unknown
Selenoprotein V	SelV	Unknown	Unknown

Table 1.
Types of selenoproteins in humans, their names, location, and functions [23].

that increased selenoprotein activity in vascular endothelial cells suppressed adhesion induced by a proinflammatory cytokine [34, 35]. In addition, it has been determined that selenoproteins protect the structure of the vessel wall by dissolving the cholesterol accumulated in the blood vessel wall [36]. In another study, it was reported that SELENOS has preventive effects on atherosclerosis and hypertension [20].

5. The function of selenoproteins in inhibiting ER stress

The endoplasmic reticulum is an organelle in the eukaryotic cell that spreads throughout the cell, especially involved in protein synthesis. When the ER is opened too much, the ER stress response occurs due to misfolded proteins and imbalances in calcium homeostasis. This causes cell apoptosis [34]. Some selenoproteins, SELENON, SELENOK, SELENOM, specifically the 15 kDa selenoproteins DIO2, SELENOS, and SELENOT, regulate ER stress [35–38]. Selenoproteins located in the ER is involved in regulating oxidative stress, inflammation, and intracellular Ca homeostasis. SELENON acts as a cofactor for the ryanodine receptor on the ER membrane and thus regulates the intracellular Ca level [20], while Sep15 is also involved in protein folding [39]. Aforesaid, GPx1 can reduce the accumulation of proinflammatory factors and increase the body's antioxidant capacity and expression [40]. It is affected by the Nrf2/ARE pathway [41]. When the body is exposed to oxidative stress, Nrf2 dissociates from the Keap1 protein, enters the nucleus, and binds to ARE, activating the Nrf2/ARE pathway, enhancing downstream GPx1 gene expression, and attenuating oxidative stress [42, 43]. Selenoprotein expression can reduce the expression of inflammatory factors and attenuate the NO-induced proinflammatory response [38]. NADPH oxidase (NOX) can mediate excessive ROS production [44, 45], thereby suppressing ER stress that oxidative stress induces. In addition, selenoproteins increase the enzyme level of DNA methyltransferase 1 (DNMT1) and protect the cell against oxidative stress and ER stress [46] (**Figure 2**).

6. Function of selenoproteins in various diseases

In various studies, it has been reported that there are some differences in selenoprotein types and levels in some diseases. Selenium deficiency causes muscle disorders in humans and animals. White muscle disease is a disease in animals characterized by a selenium deficiency. In this disease, skeletal and cardiac muscles show white streaks due to calcium deposition. White muscle disease can affect both the skeletal and cardiac muscles in which SelW is highly expressed [21]. SelW derives its name from white muscle disease, and SelW levels are upregulated in muscle cells in response to

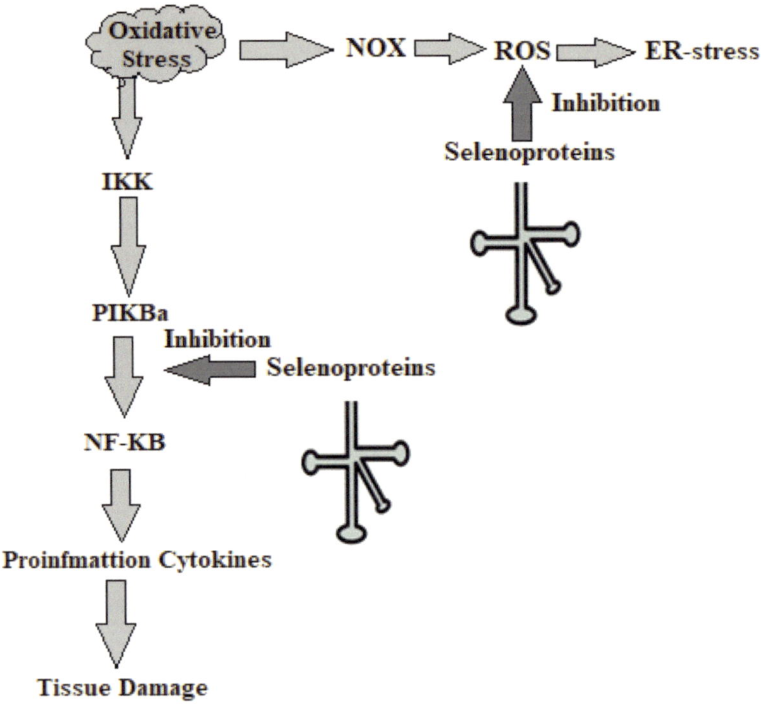

Figure 2.
The effects of selenoproteins in ER stress and inflammation [40].

exogenous oxidants [47, 48]. In the case of oxidative stress, damage to vascular endothelial cells occurs, in which case atherosclerosis, hypertension, and congestive heart failure are exacerbated [49]. Selenoproteins prevent the progression of damage due to their antioxidant properties in cardiovascular system diseases [50]. As a result of various studies, selenium supplementation increases the expression and the activity of GPX1, GPX4, and TRXR1, thus protecting the cardiovascular system against oxidative damage [51, 52]. Various studies have shown that selenoproteins have important roles in cancer [53]. Many selenoproteins have been reported to be associated with various types of cancer. For example, polymorphisms of GPX1 have been associated with various types of cancer, including breast, prostate, lung, head, and neck cancer [54, 55]. Polymorphisms in GPX2, GPX4, and SelP have been associated with colorectal cancer, Sep15 polymorphisms with lung, SelS promoter polymorphisms with stomach, and SelP polymorphisms with prostate cancer [56–59]. Studies have shown that selenoproteins play an important role in preventing neurological disorders. Some of the dietary selenium is stored in the brain tissue and it has been determined that it has a protective effect on the brain tissue in nervous system diseases such as ROS-induced Alzheimer's, Parkinson's, and ischemic brain damage [60–62]. In some studies, it has been determined that selenoproteins are protective against hyperglycemia-induced increased ROS production and resulting tissue damage in diabetes mellitus [63, 64].

7. Hazards of selenium supplementation

Apart from these mentioned issues, excessive intake of selenium causes harmful effects on the organism. If selenium absorption is excessive, selenium excess, in other words, selenium poisoning, selenium toxicity, or selenosis occur [65]. In the case of selenosis, mood changes are seen due to fatigue, vomiting, diarrhea, changes in nail structure, hair loss, or nerve damage [66]. In addition, excess selenium can cause such severe damage to the liver or heart tissue that they cannot adequately perform their liver and heart functions [67]. In case of damage to the liver tissue to this extent, cirrhosis, heart failure, which leads to damage to the heart and deterioration of heart functions, occurs [68]. When selenium comes into contact with the skin and mucous membranes, it also damages these organs [69]. Damage to the skin and mucous membranes is manifested, among other signs, by skin blistering. Excess selenium in the organism may lead to the development of malignant tumors other than those listed above [70]. For this reason, selenium in the composition of cigarettes is thought to cause cancer.

8. Conclusion

Selenium shows its effect on living things through selenoproteins. Its main biological form is selenocysteine, and its synthesis begins with the binding of the serine amino acid to tRNA. Selenocysteine is similar to cysteine, but it has a selenium atom instead of sulfur and is ionized at physiological pH. In the study, replacing selenocysteine with cysteine significantly reduces enzyme activity. This supports the critical role of the ionized selenium atom. Selenoproteins contain one or more selenocysteine residues in their primary structure. Selenium can enter the body in various ways, but its absorption is mainly in the form of selenoprotein. As a result of various studies, 25 selenoproteins, 5 of which are glutathione, have been isolated in humans. These selenoproteins are GPXs, TrxRs, DIOs, SPS, SELENOS, SELENOO, SELENOT, SELENOH, SELENOP, SELENOW, SELENOM, SELENON, SELENOI, SELENOC, 15 kDa, SELENOR, and SELENOV. Selenoproteins are found in various parts of the cell such as mitochondria, endoplasmic reticulum, nucleus, cell membrane, and Golgi membrane. And where they are found, they have various functions such as antioxidant, anti-inflammatory, hormone metabolism, and regulation of ER stress. In this study, the synthesis, types, locations, and roles of cell proteins in inflammation and ER stress are explained.

Author details

Volkan Gelen[1*], Adem Kara[2] and Abdulsamed Kükürt[3]

1 Faculty of Veterinary Medicine, Department of Physiology, Kafkas University, Kars, Turkey

2 Faculty of Science, Department of Genetics, Erzurum Technical University, Erzurum, Turkey

3 Faculty of Veterinary Medicine, Department of Biochemistry, Kafkas University, Kars, Turkey

*Address all correspondence to: gelen_volkan@hotmail.com

References

[1] Rayman MP. Selenium and human health. Lancet. 2012;**379**:1256-1268

[2] Rayman MP. The importance of selenium to human health. Lancet. 2000;**356**:233-241

[3] Kieliszek M. Selenium(−)fascinating microelement, properties and sources in food. Molecules. 2019;**24**:1298

[4] Dinh QT, Cui Z, Liang D. Selenium distribution in the Chinese environment and its relationship with human health: A review. Environment International. 2018;**112**:294-309

[5] Liu X, He S, Tan W. Expression profile analysis of selenium-related genes in peripheral blood mononuclear cells of patients with Keshan disease. BioMed Research International. 2019;**2019**:4352905

[6] Wang K, Yu J, Sun D. Endemic Kashin-Beck disease: A food-sourced osteoarthropathy. Seminars in Arthritis and Rheumatism. 2020;**50**:366-372

[7] Wang L, Yin J, Guo X. Serious selenium deficiency in the serum of patients with Kashin-beck disease and the effect of nano-selenium on their chondrocytes. Biological Trace Element Research. 2020;**194**:96-104

[8] Hadrup N, Ravn-Haren G. Absorption, distribution, metabolism and excretion (ADME) of oral selenium from organic and inorganic sources: A review. Journal of Trace Elements in Medicine and Biology. 2021;**67**:126801

[9] Mehdi Y, Hornick JL, Dufrasne I. Selenium in the environment, metabolism and involvement in body functions. Molecules. 2013;**8**:3292-3311

[10] Lu J, Holmgren A. Selenoproteins. The Journal of Biological Chemistry. 2009;**284**:723-727

[11] Labunskyy VM, Hatfield DL, Gladyshev VN. Selenoproteins: Molecular pathways and physiological roles. Physiological Reviews. 2014;**94**:739-777

[12] Zeng R, Farooq MU, Zhu J. Dissecting the potential of selenoproteins extracted from selenium-enriched rice on physiological, biochemical and anti-ageing effects in vivo. Biological Trace Element Research. 2020;**196**:119-130

[13] Xu X, Bao Y, Wu J. Chemical analysis and flavor properties of blended orange, carrot, apple and Chinese jujube juice fermented by selenium-enriched probiotics. Food Chemistry. 2019;**289**:250-258

[14] Adadi P, Barakova NV, Krivoshapkina EF. Designing selenium functional foods and beverages: A review. Food Research International. 2019;**120**:708-725

[15] Hoffmann PR, Berry MJ. Selenoprotein synthesis: A unique translational mechanism used by a diverse family of proteins. Thyroid. 2005;**15**:769-775

[16] Bulteau AL, Chavatte L. Update on selenoprotein biosynthesis. Antioxidants & Redox Signaling. 2015;**23**:775-794

[17] Santesmasses D, Mariotti M, Gladyshev VN. Bioinformatics of selenoproteins. Antioxidants & Redox Signaling. 2020;**33**:525-536

[18] Sunde RA, Raines AM. Selenium regulation of the selenoprotein and nonselenoprotein transcriptomes

in rodents. Advances in Nutrition. 2011;**2**:138-150

[19] Metanis N, Hilvert D. Natural and synthetic selenoproteins. Current Opinion in Chemical Biology. 2014;**22**:27-34

[20] Hariharan S, Dharmaraj S. Selenium, and selenoproteins: it's role in the regulation of inflammation. Inflammopharmacology. 2020;**28**:667-695. DOI: 10.1007/s10787-020-00690-x

[21] Papp LV, Lu J, Khanna KK. From selenium to selenoproteins: Synthesis, identity, and their role in human health. Antioxidants & Redox Signaling. 2007;**9**:775-806

[22] Schoenmakers E, Chatterjee K. Human disorders affecting the selenocysteine incorporation pathway cause systemic selenoprotein deficiency. Antioxidants & Redox Signaling. 2020;**33**:481-497

[23] Meplan C, Hesketh J. The infl uence of selenium and selenoprotein gene variants on colorectal cancer risk. Mutagenesis. 2012;**27**(2):177-186

[24] Gelen V, Şengül E, Yıldırım S, et al. The protective effects of hesperidin and curcumin on 5-fluorouracil–induced nephrotoxicity in mice. Environmental Science and Pollution Research. 2021;**28**:47046-47055

[25] Gelen V, Şengül E, Yıldırım S, Atila G. The protective effects of naringin against 5-fluorouracil-induced hepatotoxicity and nephrotoxicity in rats. Iranian Journal of Basic Medical Sciences. 2018;**21**(4):404-410

[26] Arbogast S, Ferreiro A. Selenoproteins and protection against oxidative stress: Selenoprotein N as a novel player at the crossroads of redox signaling and calcium homeostasis. Antioxidants & Redox Signaling. 2010;**12**:893-904

[27] Gelen V, Yıldırım S, Şengül E, Çınar A, Çelebi F, Küçükkalem M, et al. Naringin attenuates oxidative stress, inflammation, apoptosis, and oxidative DNA damage in acrylamide-induced nephrotoxicity in rats. Asian Pacific Journal of Tropical Biomedicine. 2022;**12**:223-232

[28] Sengul E, Gelen V, Yildirim S, Cinar İ, Aksu EH. Effects of naringin on oxidative stress, inflammation, some reproductive parameters, and apoptosis in acrylamide-induced testis toxicity in rat. Environmental Toxicology. 2023;**38**(4):798-808

[29] Sengul E, Gelen V, Yildirim S, Tekin S, Dag Y. The effects of selenium in acrylamide-induced nephrotoxicity in rats: Roles of oxidative stress, inflammation, apoptosis, and DNA damage. Biological Trace Element Research. 2021;**199**(1):173-184

[30] Steven ER, Kim BW, Chu FF. The Gdac1 locus modifies spontaneous and salmonella-induced colitis in mice deficient in either Gpx2 or Gpx1 gene. Free Radical Biology & Medicine. 2013;**65**:1273-1283

[31] Kara A, Gedikli S, Sengul E, Gelen V, Ozkanlar S. Oxidative Stress and Autophagy. 1st ed. London: InTechOpen, Free Radicals and Diseases; 2016. pp. 69-86

[32] Reszka E. Selenoproteins in bladder cancer. Clinica Chimica Acta. 2012;**413**:847-854

[33] Yang Z, Liu C, Li S. Selenium deficiency mainly influences antioxidant selenoproteins expression in broiler immune organs. Biological Trace Element Research. 2016;**172**:209-221

[34] Rees K, Hartley L, Stranges S. Selenium supplementation for the primary prevention of cardiovascular disease. Cochrane Database of Systematic Reviews. 2013;**31**:1

[35] Benstoem C, Goetzenich A, Stoppe C. Selenium and its supplementation in cardiovascular disease—What do we know? Nutrients. 2015;7:3094-3118

[36] Shalihat A, Hasanah AN, Gozali D. The role of selenium in cell survival and its correlation with protective effects against cardiovascular disease: A literature review. Biomedicine & Pharmacotherapy. 2021;**134**:111125

[37] Chi Q, Zhang Q, Li S. Roles of selenoprotein S in reactive oxygen species-dependent neutrophil extracellular trap formation induced by selenium-deficient arteritis. Redox Biology. 2021;**44**:102003

[38] Rocca C, Boukhzar L, Angelone T. A selenoprotein T-derived peptide protects the heart against ischaemia/reperfusion injury through inhibition of apoptosis and oxidative stress. Acta Physiologica. 2018;**223**:e13067

[39] Nettleford SK, Zhao L, Tsuji PA. Selenium and the 15kDa selenoprotein impact colorectal tumorigenesis by modulating intestinal barrier integrity. International Journal of Molecular Sciences. 2021;**22**:10651

[40] Gelen V, Şengül E, Gedikli S, Gür C, Özkanlar S. Therapeutic effect of quercetin on renal function and tissue damage in the obesity-induced rats. Biomedicine & Pharmacotherapy. 2017;**89**:524-528

[41] Kükürt A, Karapehlivan M, Gelen V. The use of Astaxanthin as a natural antioxidant on ovarian damage. In: Karapehlivan M, Kükürt A, Gelen V, editors. Animal Models and Experimental Research in Medicine. London: IntechOpen; 2022. DOI: 10.5772/intechopen.108854

[42] Kükürt A, Gelen V, Başer ÖF, Deveci HA, Karapehlivan M. Thiols: Role in oxidative stress-related disorders. In: Atukeren P, editor. Accenting Lipid Peroxidation. London: IntechOpen; 2021. pp. 27-47. DOI: 10.5772/intechopen.96682

[43] Karamese M, Guvendi B, Karamese SA, Cinar I, Can S, Erol HS, et al. The protective effects of epigallocatechin gallate on lipopolysa saccharide-induced hepatotoxicity: An in vitro study on Hep3B cells. Iranian Journal of Basic Medical Sciences. 2016;**19**(5):483-489

[44] Şengül E, Gelen V, Gedikli S, Özkanlar S, Gür C, Çelebi F, et al. The protective effect of quercetin on cyclophosphamide-induced lung toxicity in rats. Biomedicine & Pharmacotherapy. 2017;**92**:303-307

[45] Gelen V, Sengul E. Antioxidant, anti-inflammatory, and antiapoptotic effects of naringin on cardiac damage induced by cisplatin. IJTK. 2020;**19**(2):459-465

[46] Ye R, Huang J, Wang Z, Chen Y, Dong Y. The role and mechanism of essential Selenoproteins for homeostasis. Antioxidants (Basel). 2022;**11**(5):973

[47] Beilstein MA, Vendeland SC, Barofsky E, Jensen ON, Whanger PD. Selenoprotein W of rat muscle binds glutathione and an unknown small molecular weight moiety. Journal of Inorganic Biochemistry. 1996;**61**:117-124

[48] Vendeland SC, Beilstein MA, Yeh JY, Ream W, Whanger PD. Rat skeletal muscle selenoprotein W: cDNA clone and mRNA modulation by dietary selenium. Proceedings. National Academy of Sciences. United States of America. 1995;**92**:8749-8753

[49] Lum H, Roebuck KA. Oxidant stress and endothelial cell dysfunction. American Journal of Physiology. Cell Physiology. 2001;**280**:C719-C741

[50] Miller S, Walker SW, Arthur JR, Nicol F, Pickard K, Lewin MH, et al. Selenite protects human endothelial cells from oxidative damage and induces thioredoxin reductase. Clinical Science. 2001;**100**:543-550

[51] Steinbrenner H, Alili L, Bilgic E, Sies H, Brenneisen P. Involvement of selenoprotein P in the protection of human astrocytes from oxidative damage. Free Radical Biology & Medicine. 2006;**40**:1513-1523

[52] Tang R, Liu H, Wang T, Huang K. Mechanisms of selenium inhibition of cell apoptosis induced by oxysterols in rat vascular smooth muscle cells. Archives of Biochemistry and Biophysics. 2005;**441**:16-24

[53] Foster CB, Aswath K, Chanock SJ, McKay HF, Peters U. Polymorphism analysis of six selenoprotein genes: Support for a selective sweep at the glutathione peroxidase 1 locus (3p21) in Asian populations. BMC Genetics. 2006;**7**:56

[54] Hu Y, Benya RV, Carroll RE, Diamond AM. Allelic loss of the gene for the GPX1 selenium-containing protein is a common event in cancer. The Journal of Nutrition. 2005;**135**:3021S-3024S

[55] Hu YJ, Diamond AM. Role of glutathione peroxidase 1 in breast cancer: Loss of heterozygosity and allelic differences in the response to selenium. Cancer Research. 2003;**63**:3347-3351

[56] Al-Taie OH, Uceyler N, Eubner U, Jakob F, Mork H, Scheurlen M, et al. Expression profiling and genetic alterations of the selenoproteins GI-GPx and SePP in colorectal carcinogenesis. Nutrition and Cancer. 2004;**48**:6-14

[57] Bermano G, Pagmantidis V, Holloway N, Kadri S, Mowat NA, Shiel RS, et al. Evidence that a polymorphism within the 3′UTR of glutathione peroxidase 4 is functional and is associated with susceptibility to colorectal cancer. Genes & Nutrition. 2007;**2**:225-232

[58] Jablonska E, Gromadzinska J, Sobala W, Reszka E, Wasowicz W. Lung cancer risk associated with selenium status is modified in smoking individuals by Sep15 polymorphism. European Journal of Nutrition. 2008;**47**:47-54

[59] Cooper ML, Adami HO, Gronberg H, Wiklund F, Green FR, Rayman MP. Interaction between single nucleotide polymorphisms in selenoprotein P and mitochondrial superoxide dismutase determines prostate cancer risk. Cancer Research. 2008;**68**:10171-10177

[60] Behne D, Hilmert H, Scheid S, Gessner H, Elger W. Evidence for specific selenium target tissues and new biologically important selenoproteins. Biochimica et Biophysica Acta. 1988;**966**:12-21

[61] Nakayama A, Hill KE, Austin LM, Motley AK, Burk RF. All regions of mouse brain are dependent on selenoprotein P for maintenance of selenium. The Journal of Nutrition. 2007;**137**:690-693

[62] Chen J, Berry MJ. Selenium and selenoproteins in the brain and brain diseases. Journal of Neurochemistry. 2003;**86**:1-12

[63] Aydemir-Koksoy A, Turan B. Selenium inhibits proliferation signaling and restores sodium/potassium pump function of diabetic rat aorta.

Biological Trace Element Research. 2008;**126**:237-245

[64] Ozdemir S, Ayaz M, Can B, Turan B. Effect of selenite treatment on ultrastructural changes in experimental diabetic rat bones. Biological Trace Element Research. 2005;**107**:167-179

[65] Lv Q, Liang X, Nong K, Gong Z, Qin T, Qin X, et al. Advances in research on the toxicological effects of selenium. Bulletin of Environmental Contamination and Toxicology. May 2021;**106**(5):715-726

[66] Poos MI, Vorosmarti AL, Ramsey MR. Institute of Medicine (IOM) Dietary Reference Intakes for Vitamin C, Vitamin E, Selenium, and Carotenoids. Washington, DC: National Academy Press; 2000. Selenium. pp. 284-324

[67] Levander OA. Scientific rationale for the 1989 recommended dietary allowance for selenium. Journal of the American Dietetic Association. 1991;**91**(12):1572-1576

[68] Fan AM, Kizer KW. Selenium-nutritional, toxicologic, and clinical aspects. The Western Journal of Medicine. 1990;**153**(2):160-167

[69] Nuttall KL. Evaluating selenium poisoning. Annals of Clinical and Laboratory Science. 2006;**36**(4):409-420

[70] Yang GQ, Wang SZ, Zhou RH, Sun SZ. Endemic selenium intoxication of humans in China. The American Journal of Clinical Nutrition. 1983;**37**(5):872-881

Chapter 2

An Overview of the Antioxidant and Anti-Inflammatory Activity of Selenium

Mehmet Başeğmez

Abstract

Selenium, whose name comes from the Greek word for "Selene," has been a topic of interest as a micronutrient ever since it was described in 1817 as a by-product of sulfuric acid manufacturing. Selenium, the most important micronutrient for both humans and animals, must be consumed daily to support the body's natural metabolism and homeostasis. The small intestine is responsible for the absorption of selenium in both its organic and inorganic forms. Selenium is then able to be widely distributed throughout the body's diverse tissues, where it plays an important role in the regulation of the synthesis of selenoproteins. The synthesis of human selenoproteins involves the incorporation of a selenium-containing homolog of cysteine in each of the 25 selenium-containing proteins that make up this series. Many selenoproteins, including glutathione peroxidase (GPX), thioredoxin reductase (TrxR), and iodothyronine deiodinases (IDD), function as crucial cellular defenses against oxidative stress. Therefore, selenium is extremely important in boosting antioxidant defense. Recent studies have also shown that there is a close relationship between selenium and inflammation, and that selenium has regulatory effects on inflammation by affecting the expression of various cytokines. This chapter's goal was to thoroughly review the research on how selenium is related to antioxidant and anti-inflammatory activity.

Keywords: antioxidant, anti-inflammatory, human nutrition, selenium, selenoproteins

1. Introduction

Selenium, which takes its name from the Greek word "Selene," has been attracting attention as a trace element since 1817 as a by-product of sulfuric acid [1]. Both environmental and endogenous factors affect body selenium homeostasis [2]. Selenium can be absorbed by the small intestine in both organic and inorganic forms, after which it can be distributed throughout the body and perform important biological functions, most particularly by controlling the synthesis of selenoproteins [3]. Selenoproteins play an important role in many biochemical and physiological processes in both humans and animals because of their antioxidant properties [4]. They have antioxidant and anti-inflammatory properties that help to regulate immune cell functions [5]. Twenty-five genes in the human genome are responsible for the coding of selenocysteine-containing proteins. The selenoprotein family, whose functions are known, is named according to these functions: glutathione peroxidases (GPX1, GPX2, GPX3, GPX4, and GPX6), thioredoxin reductases (TrxR1–3), iodothyronine deiodinases (DIO1–3), selenophosphate synthetase 2 (SEPHS2), methionine sulfoxide reductase B1(MSRB1), SEP15 (SELENOF), SELH (SELENOH), SELI (SELENOI), SELK (SELENOK), SELM (SELENOM), SELN (SELENON), SELO (SELENOO), SELP (SELENOP), SELS (SELENOS), SELT (SELENOT), SELV (SELENOV), and SELW (SELENOW) [6]. The primary function of multiple selenoproteins is to protect cells from oxidative damage by taking action as major antioxidants.

In this review, I want to show how selenium affects many biological effects, mostly through selenoproteins, as well as how it affects the physiological and biochemical processes it interacts with. Furthermore, the effect of deficiency and excess selenium in the body on the antioxidant and anti-inflammatory systems and the most recent findings on human health are highlighted.

1.1 Selenium requirement in the human body

Selenium is a crucial trace element required for the proper working of all organisms. It is emphasized that very high and very low selenium levels in humans are harmful to health [7]. For instance, not getting sufficient selenium can cause oxidative stress, which decreases the concentrations of selenoproteins, such as GPx and TXNRD, in the body. On the other hand, too much selenium can cause oxidative stress by oxidizing and cross-linking protein thiol groups, which causes reactive oxygen species to form [8]. The amount of this element, which varies according to bioavailability, geographical region, and nutrition, plays an important role in selenium homeostasis in the organism. It has been determined that 40–70 micrograms [9] of this element is optimal for normal biochemical and physiological processes [10, 11]. The World Health Organization suggests that adults consume 55 µg of selenium per day [12]. The US Food and Nutrition Board determined it to be 40–70 µg for men and 45–55 µg for women [13–15]. The determination of reference values for selenium in adults is based on saturation of the plasma selenoprotein P (SePP) level with adequate selenium intake. SePP saturation was reached in people with an average body weight of 58 kg who lived in areas with low selenium levels by giving them 49 microgram of selenium every day [16]. This is equivalent to getting about 1 micrograms of selenium per kilogram of body weight every day [17]. Reference values for children and teens are based on values made for adults, with their body weight and growth factors taken into account. Estimated values for selenium intake by age groups and body weights are as follows: 15 µg/day for 1 to 4 years old, 20 µg/day for 4–7 years old, 30 µg/day for 7 to 10 years old, 45 µg/day for 10 to 13 years old, and 60 µg/day for 13 to 15 years old. The estimated daily value of selenium intake for boys aged 15 to 19 is 70 micrograms, while for girls of the same age, it is 60 micrograms Daily [17]. The determination of selenium requirements in newborns and 4-month-old infants is based on the selenium content of breast milk [17]. A daily average of 750 ml of breast milk [18] results in a selenium intake of nearly 11 µg/day. An estimate of optimal selenium intake for breastfed infants between new-born and 4 months of age is 10 micrograms. However, considering the average body weight differences and solid food intake processes in infants aged 4–12 months, an estimated daily 15 micrograms was determined for infants (**Table 1**) [17].

	Selenium µg/day		References
Age	**Male (µg)**	**Female (µg)**	
Birth–4 months	10		
4–12 months	15		
1–4 years	15		
4–7 years	20		
7–10 years	30		
10–13 years	45		[17, 19]
13–15 years	60		
15–19 years	70	60	
19–25 years	70	60	
25–51 years	70	60	
51–65 years	70	60	
Pregnant women		60	
Lactating women		75	

Table 1.
Values predicted to ensure sufficient selenium consumption.

1.2 Source of selenium in the human body

Selenium is mostly orally taken into the human organism. Plant and animal products are the main sources of this element. Selenium can be found in foods and biological materials as inorganic compounds, as well as organic compounds [20, 21]. Plants store selenium in the form of inorganic compounds called selenate (IV) or (VI) and then convert them into organic forms such as selenomethionine and selenocysteine [7]. Selenocysteine levels are high in animal-derived products [22]. Selenium is found in low concentrations in vegetables and fruits, but in high concentrations in seafood, grains, and meat products [23, 24]. On the other hand, protein-rich foods contain higher levels of selenium than foods low in protein [7]. Cereal products provide approximately 50% of the daily selenium intake, while meat, fish, and poultry products provide approximately 35%. Water and beverage products provide about 5–25% of selenium. Fruit, on the other hand, meets about 10% of the selenium demand (**Table 2**).

2. The role of selenium in oxidative stress, inflammation, and immunity

Oxidative stress is a disruption of the balance between the prooxidant and antioxidant systems in the body [27, 28]. In normal circumstances, the prooxidant system and the antioxidant system work together to maintain the body's homeostasis. However, increased prooxidant system activity and deterioration of the antioxidant system (**Table 3**) result in oxidative stress. The development of many chronic diseases, including diabetes [30], cancer [31], antiviral agents [32], and various aging-related and central nervous system (CNS) disorders [33], can result in high levels of reactive oxygen and nitrogen species production. In addition, reactive oxygen

Selenium Source	Food	Selenium concentration(mg/kg)	Selenium forms	References
Meat and meat products	Beef	0.042–0.142	Selenomethionine	[19, 25]
	Lamp	0.033–0.260		
	Chicken	0.081–0.142	Selenomethionine/ Selenocysteine	
	Pork	0.032–0.198	Selenomethionine/Selenate	
	Fish	0.1–5.0	Selenomethionine/Selenite/ Selenate	[19]
Milk and dairy products	Milk	0.01–0.03	Selenocysteine/Selenite	
Vegetable products	Broccoli	0.5–1.0	Selenomethionine/Selenate	
	Garlic	0.05–1.0	Selenomethionine/ Selenocysteine	
	Potatoes	0.12	Selenomethionine	[19, 25]
	Mushrooms	0.01–1.40	Selenomethionine/ Selenocysteine/ Selenomethylselenocysteine	[26]
	Onions	0.02–0.05	Selenomethionine/ Selenocysteine	[19]
Grain products	Bread	0.01–30	Selenomethionine/Selenate	
	Cereal	0.02–35		
	Lentils	0.24–0.36		[19, 25]
	Rice	0.05–0.08	Selenomethionine	
Other food products	Yeast	0.6–15		[19]
	Eggs	3–25	Selenomethionine/ Selenocysteine	

Table 2.
Selenium concentrations in various foods.

production causes intense lipid peroxidation in cells, causing the breakdown of cell membranes [5]. As a result, cellular homeostasis is disrupted, and human health is affected. Antioxidant activity as a free radical scavenger is linked to protecting cells from autooxidation and keeping their structure so that the immune system can work at its best [34].

In the process of regulating antioxidant activities, various selenoproteins are essential players [35]. Glutathione peroxidase GSH-Px, which contains one selenium atom in each subunit, was one of the first highly effective selenoproteins [36]. The glutathione peroxidase enzyme reduces reactive oxygen and nitrogen species by converting hydrogen peroxide (H_2O_2) to water (H_2O) and organic hydroperoxides (ROOH) to alcohol (ROH) [14, 37]. The selenium dependent (GPXs 1–4) significantly detoxifies cellular peroxides that protect against reactive oxygen species [38]. Glutathione peroxidase 1 (GPX1) is the most common selenoprotein that protects the body from oxidative stress caused by reactive oxygen and nitrogen [39]. On the other

	Radicals	Non- Radicals
Reactive oxygen species	O_2.–, Superoxide	H_2O_2, Hydrogen peroxide
	OH., Hydroxyl	HOCI, Hypochlorous acid
		$^1O_2$, Singlet oxygen
	RO_2., Peroxyl	O_3, Ozone
	RO., Alkoxyl	
	HO_2., Hydroperoxyl	
	NO., Nitric oxide	
	NO_2., Nitrogen dioxide	
Reactive nitrogen species	NO., Nitric oxide	HNO_2, Nitrous acid
	NO_2., Nitrogen dioxide	N_2O_4, Dinitrogen tetroxide
		N_2O_3, Dinitrogen trioxide
		ONOO-, Peroxynitrite
		ONOOH, Peroxynitrous acid
		NO_2+, Nitronium cation
		ROONO, Alkyl peroxynitrites

Table 3.
Reactive oxygen and nitrogen species [29].

hand, GPX1 may also decrease the concentration of lipid hydroperoxides and other hydroperoxides once they have been released from membrane lipids [40]. In the same way, as GPX1 does, GPX2 neutralizes H_2O_2 and fatty acid hydroperoxides [41]. This selenoprotein, which was expressed in the intestinal tract in the early 1990s, has also attracted attention with its antioxidant activities by affecting apoptosis and regulating the self-renewal of the intestinal epithelium [42]. GPX3, found in plasma and milk [38], is an important selenoprotein that serves as a source of extracellular antioxidant capacity, especially in the kidney proximal tubule epithelial cell [43], by reducing oxidative stress in the heart, liver, lungs, skeletal muscle, and thyroid gland [44, 45]. GPX4 is unique among GPXs in that it has the ability to catalyze the reduction of hydrogen peroxide and other lipid hydroperoxides in addition to reducing phospholipid hydroperoxides [46]. GPX6 enzyme expression was detected only in the embryo and olfactory epithelium [47]. In an *in vivo* study, supplementation of selenium-rich, rice-extracted selenoproteins to male mice modeled aging by abdominal D-galactose injection and increased GSH-Px and superoxide dismutase (SOD) enzyme activation in the liver and serum of mice compared to the control group [48]. TrxR enzymes, which function in concert with NADPH to clear the redox system in mammals, have been identified in three different forms [49]. Trx1 is responsible for the reduction of thioredoxins in the cytosol, TrxR2 for the reduction of thioredoxins in the mitochondria, and TrxR3 for the reduction of glutathione and glutaredoxin [50]. DNA synthesis, which occurs at the beginning of cellular processes, relies on the existence of selenium in the catalytic region of TrxR [51]. Furthermore, mammalian TrxRs are selenoproteins that play an essential function in many cellular processes by modulating the action of the core redox molecule thioredoxin, as well as directly reducing a variety of substrates [50]. DIOs are members of the selenoprotein family that include the three enzymes (DIO1, DIO2, and DIO3) that catalyze the activation (DIO1) and inactivation (DIO2) of the thyroid hormone thyroxine (T4), respectively [52]. DIO1 is involved in T3 production in the thyroid gland and controlling circulating T3 levels, while DIO2 and DIO3 are involved in local deiodination processing processes at the tissue and organ level [53]. Increased oxidative damage in thyroid tissue has been associated with decreased DIO and GPx activity in the organism and insufficient GPx concentration [54]. In mammals, selenophosphate synthetase 2 (SEPHS2) is a selenoprotein involved in the biosynthesis of the amino acid selenocysteine, which catalyzes the formation of selenophosphate from selenide and ATP [55, 56]. SelR, commonly referred to as methionine-R-sulfoxide reductase B1 (MsrB1), is a protein that helps reduce oxidized methionine (Met) residues (methionine sulfoxides) [57]. SelR comprises a redox effective selenoprotein containing a particular enzymatic activity that is necessary for oxidative protein repair [50]. SEP15 is the first selenoprotein [58] to be widely distributed across multiple organs including the brain, lung, testis, liver, thyroid, and kidney [59]. Sep15, belonging to the class of thiol-disulfide oxidoreductase-like selenoproteins [60], is a selenoprotein exhibiting redox activity [61]. Selenoprotein K is mainly expressed in the heart and skeletal muscle, but it is also found in other tissues such as the placenta, liver, and pancreas. Increasing levels of SELK in the organism exhibit antioxidant properties in the heart by reducing intracellular ROS levels and protecting cardiomyocytes against oxidative damage [62]. Selenoprotein M, a selenoprotein distantly related to Sep15, acts as a redox regulator with the amino acid selenocysteine [63]. SELM, induced by sodium selenite, which has prooxidant properties, has a functional role in catalyzing free radicals [64]. SELN, which is an endoplasmic reticulum glycoprotein and has important functions in muscle tissue, has been associated with myopathies [65]. SELN, which draws attention with its cell proliferation and regeneration, is significantly effective in the early embryonic development process [66]. It plays an important role in the redox system by contributing to calcium homeostasis in the organism [67] and protecting the cells from oxidative stress [68]. SelO, which is located in the mitochondria of the organism and draws attention with its feature of being the biggest selenoprotein [69], plays a role in oxidative stress by controlling S-glutathionylation levels [70]. Selenoprotein P is estimated to contain 50% of plasma selenium [71]. The plasma concentration of SELP varies depending on selenium supplementation. These changes in selenium intake, together with its concentration at the plasma SELP level, may reflect an indication of the amino acid protein residues of selenite in its molecule [72]. SELP, which exhibits antioxidant properties, has been shown to protect astrocytes [73] and endothelial cells from oxidative stress [74, 75]. In addition, it has been demonstrated that SELP prevents the oxidation of low-density lipoproteins [76]. Selenoprotein S, one of the resident proteins of the endoplasmic reticulum, is a selenoprotein involved in the reduction of reactive oxygen species and redox signaling [77]. This selenoprotein plays critical functions in protein quality control processes, cytokine modulation, and signaling [78]. Selenoprotein T is the only protein among the selenoproteins located in the membrane of the endoplasmic reticulum. The decrease in the expression of selenoprotein T, known for its suppressive effect on reactive oxygen and nitrogen species, has been shown as a possible factor in the deterioration of the antioxidant balance [79]. Selenoprotein V, which is predominantly localized in the intracellular cytoplasm, plays an important role, such as other selenoproteins, in the elimination of oxidative stress by protecting against endoplasmic reticulum stress and apoptosis caused by prooxidants [80]. Selenoprotein W, which is expressed in every tissue, is one of the well-known selenoproteins with antioxidant properties that are very important for the proper growth of the brain and embryo [81, 82].

Selenium, which plays an important role in antioxidant defense for body homeostasis, also plays an important role in the regulation of different inflammatory processes in the organism [83]. Adequate selenium supplements are essential for the immune system. For example, selenoprotein expression is affected in male mice supplemented with selenium, and immune response pathways, such as Interferon-γ and IL-6, are supported [84]. Interleukin IL-2, IL-4, IL-5, IL-13, and IL-22 cytokine levels were significantly higher in plasma and peripheral blood mononuclear cells in people who ate 200 mg of selenium-rich broccoli per serving for three days [85]. A previous study showed that increasing selenium supplements increased antigen-specific $CD4^+$ T cell responses. In addition, high selenium diets increased interferon-gamma (IFN-γ) and IL-2 expression levels compared to low and moderate selenium diets [86]. The higher contents of selenium in the blood of older individuals have been shown to have a positive correlation with a higher percentage and activity of natural killer (NK) cells [87]. In patients with acute respiratory distress syndrome, intravenous selenium supplementation attenuated inflammatory responses and significantly improved respiration by restoring the antioxidant capacity of the lungs *via* IL-1β and IL-6 proinflammatory cytokine levels [88]. Selenium supplementation significantly affects both innate immunity (neutrophils, macrophages, and NK) and acquired immunity (T and B lymphocytes) [89]. The phagocytosis functions of macrophages and the T cell activities of the body were significantly boosted by selenium-containing proteins [90]. Selenoprotein K plays an important role in the regulation of immunity by affecting the proliferation of T cells and the transport of neutrophils as a cofactor for the enzyme involved in the maturation of proteins in the endoplasmic reticulum to support calcium influx [5, 91].

3. Conclusions

These findings suggest that adequate selenium supplements may contribute to the body's immune homeostasis. It also shows that the selenoprotein family can prevent damage to cellular proteins by directly scavenging reactive oxygen and nitrogen species. In this respect, selenium appears to have both a protective and a therapeutic role in immune dysfunction, and further research is needed to understand the effect of selenium at different pharmacological doses, different administration methods, and in different age and gender groups. However, with new studies to be done, it is necessary to reveal the mechanisms that play a role in selenium homeostasis depending on oral or parenteral supplements in humans and animals. In addition, due to the fact that the drugs used in the treatment of chronic diseases all over the world, including in our country, have both side effects and are expensive, it leads to an increase in health costs and causes countries to determine new principles in health services. In recent years, scientists have accelerated their studies to find more accessible, inexpensive, and low side effect products such as selenium instead of expensive, prescription-only pharmacological agents with high side effects. As a result of the promising findings on the effects it creates in the organism, selenium supplements may be used as a potential pharmacological agent in the prevention of oxidative stress and regulation of inflammation in the near future.

Author details

Mehmet Başeğmez
Department of Veterinary, Acıpayam Vocational High School, Laborant and Veterinary Health Program, Pamukkale University, Denizli, Turkey

*Address all correspondence to: mehmetbsgmz@gmail.com

References

[1] Shahzad SA, Shahzad SA. General introduction on selenium. In: Novel Selenium-Mediated Rearrangements and Cyclisations. 2013. pp. 1-12

[2] Park K, Rimm E, Siscovick D, Spiegelman D, Steven Morris J, Mozaffarian D. Demographic and lifestyle factors and selenium levels in men and women in the U.S. Nutrition Research and Practice. 2011;**5**(4):357-364. DOI: 10.4162/NRP.2011.5.4.357

[3] Burk RF. Biological activity of selenium. Annual Review of Nutrition. 1983;**3**(1):53-70. DOI: 10.1146/annurev.nu.03.070183.000413

[4] EFSA Panel on Dietetic Products, Nutrition and Allergies (NDA). Scientific opinion on dietary reference values for niacin. EFSA Journal. 2014;**12**(7):3759. DOI: 10.2903/j.efsa.2014.3759

[5] Zoidis E, Seremelis I, Kontopoulos N, Danezis GP. Selenium-dependent antioxidant enzymes: Actions and properties of selenoproteins. Antioxidants. 2018;**7**(5):66. DOI: 10.3390/antiox7050066

[6] Regina BF, Gladyshev VN, Arnér ES, et al. Selenoprotein gene nomenclature. Journal of Biological Chemistry. 2016;**291**(46):24036-24040. DOI: 10.1074/jbc.M116.756155

[7] Kieliszek M. Selenium–fascinating microelement, properties and sources in food. Molecules. 2019;**24**(7). DOI: 10.3390/molecules24071298

[8] Lee KH, Jeong D. Bimodal actions of selenium essential for antioxidant and toxic pro-oxidant activities: The selenium paradox (review). Molecular Medicine Reports. 2012;**5**(2). DOI: 10.3892/mmr.2011.651

[9] Combs GF. Selenium in global food systems. British Journal of Nutrition. 2001;**85**(5). DOI: 10.1079/bjn2000280

[10] Pophaly SD, Poonam SP, Kumar H, Tomar SK, Singh R. Selenium enrichment of lactic acid bacteria and bifidobacteria: A functional food perspective. Trends in Food Science and Technology. 2014;**39**(2). DOI: 10.1016/j.tifs.2014.07.006

[11] Krohn RM, Lemaire M, Negro Silva LF, et al. High-selenium lentil diet protects against arsenic-induced atherosclerosis in a mouse model. Journal of Nutritional Biochemistry. 2016;**27**:9-15. DOI: 10.1016/j.jnutbio.2015.07.003

[12] Organization WH. Trace Elements in Human Nutrition and Health World Health Organization. Geneva: World Health Organization; 1996. Published online. ISBN 92 4 156173 4

[13] Tamari Y, Kim ES. Longitudinal study of the dietary selenium intake of exclusively breast- fed infants during early lactation in Korea and Japan. Journal of Trace Elements in Medicine and Biology. 1999;**13**(3):129-133. DOI: 10.1016/S0946-672X(99)80002-9

[14] Kieliszek M, Błazejak S. Current knowledge on the importance of selenium in food for living organisms: A review. Molecules. 2016;**21**(5). DOI: 10.3390/molecules21050609

[15] Kieliszek M, Błazejak S. Selenium: Significance, and outlook for supplementation. Nutrition. 2013;**29**(5). DOI: 10.1016/j.nut.2012.11.012

[16] Xia Y, Hill KE, Li P, et al. Optimization of selenoprotein P and other plasma selenium biomarkers for the assessment of the selenium nutritional

requirement: A placebo-controlled, double-blind study of selenomethionine supplementation in selenium-deficient Chinese subjects. American Journal of Clinical Nutrition. 2010;**92**(3). DOI: 10.3945/ajcn.2010.29642

[17] Kipp AP, Strohm D, Brigelius-Flohé R, et al. Revised reference values for selenium intake. Journal of Trace Elements in Medicine and Biology. 2015;**32**:195-199. DOI: 10.1016/j.jtemb.2015.07.005

[18] Neville MC, Keller R, Seacat J, et al. Studies in human lactation: Milk volumes in lactating women during the onset of lactation and full lactation. American Journal of Clinical Nutrition. 1988;**48**(6):1375-1386. DOI: 10.1093/ajcn/48.6.1375

[19] Hariharan S, Dharmaraj S. Selenium and selenoproteins: It's role in regulation of inflammation. Inflammopharmacology. 2020;**28**(3). DOI: 10.1007/s10787-020-00690-x

[20] Dumont E, Vanhaecke F, Cornelis R. Selenium speciation from food source to metabolites: A critical review. Analytical and Bioanalytical Chemistry. 2006;**385**(7). DOI: 10.1007/s00216-006-0529-8

[21] Lobinski R, Edmonds JS, Suzuki KT, Uden PC. Species-selective determination of selenium compounds in biological materials (technical report). Pure and Applied Chemistry. 2000;**72**. DOI: 10.1351/pac200072030447

[22] Pezzarossa B, Petruzzelli G, Petacco F, Malorgio F, Ferri T. Absorption of selenium by Lactuca sativa as affected by carboxymethylcellulose. Chemosphere. 2007;**67**(2). DOI: 10.1016/j.chemosphere.2006.09.073

[23] Tinggi U. Determination of selenium in meat products by hydride generation atomic absorption spectrophotometry. Journal of AOAC International. 1999;**82**(2). DOI: 10.1093/jaoac/82.2.364

[24] Tinggi U, Reilly C, Patterson CM. Determination of selenium in foodstuffs using spectrofluorometry and hydride generation atomic absorption spectrometry. Journal of Food Composition and Analysis. 1992;**5**(4). DOI: 10.1016/0889-1575(92)90061-N

[25] Kieliszek M, Bano I, Zare H. A comprehensive review on selenium and its effects on human health and distribution in middle eastern countries. Biological Trace Element Research. 2022;**200**(3). DOI: 10.1007/s12011-021-02716-z

[26] Reilly C. Selenium: A new entrant into the functional food arena. Trends in Food Science and Technology. 1998;**9**(3). DOI: 10.1016/S0924-2244(98)00027-2

[27] Persson T, Popescu BO, Cedazo-Minguez A. Oxidative stress in alzheimer's disease: Why did antioxidant therapy fail? Oxidative Medicine and Cellular Longevity. 2014:11. DOI: 10.1155/2014/427318

[28] Salim M, Durmuş İ, Başeğmez M, Küçükkurt İ, Eryavuz A. Effects of age on the concentrations of plasma cytokines and Lipidperoxidation in sheep. Kocatepe Veterinary Journal. Published online. 2021;**14**:37-44. DOI: 10.30607/kvj.798623

[29] Halliwell B. Antioxidants in human health and disease. Annual Review of Nutrition. 1996;**16**:33-50. DOI: 10.1146/annurev.nu.16.070196.000341

[30] Oguntibeju OO. Type 2 diabetes mellitus, oxidative stress and inflammation: Examining the links. International Journal of Physiology Pathophysiology Pharmacology. 2019;**11**(3):45-63 http://www.embase.

com/search/results?subaction=viewrecord&from=export&id=L2002518543

[31] Hayes JD, Dinkova-Kostova AT, Tew KD. Oxidative stress in cancer. Cancer Cell. 2020;**38**(2):167-197. DOI: 10.1016/j.ccell.2020.06.001

[32] Doğan MF, Kaya K, Demirel HH, Başeğmez M, Şahin Y, Çiftçi O. The effect of vitamin C supplementation on favipiravir-induced oxidative stress and proinflammatory damage in livers and kidneys of rats. Immunopharmacology and Immunotoxicology. Published online. 2023:1-6. DOI: 10.1080/08923973.2023.2181712

[33] Harman D. Free radical theory of aging. Mutation Research DNAging. 1992;**275**(3-6):257-266. DOI: 10.1016/0921-8734(92)90030-S

[34] Ang A, Pullar JM, Currie MJ, Vissers MCM. Vitamin C and immune cell function in inflammation and cancer. Biochemical Society Transactions. 2018;**46**(5):1147-1159. DOI: 10.1042/BST20180169

[35] Cai Z, Zhang J, Li H. Selenium, aging and aging-related diseases. Aging Clinical and Experimental Research. 2019;**31**(8):1035-1047. DOI: 10.1007/s40520-018-1086-7

[36] Katarzyna Z, Sobiech P, Radwińska J, Rekawek W. Effects of selenium on animal health. Journal of Elementology. 2013;**18**(2):329-340. DOI: 10.5601/jelem.2013.18.2.12

[37] Bjørklund G, Shanaida M, Lysiuk R, et al. Selenium: An antioxidant with a critical role in anti-aging. Molecules. 2022;**27**(19):6613. DOI: 10.3390/molecules27196613

[38] Antonyak H, Iskra R, Panas N, Lysiuk R. Selenium. In: Healthy Ageing and Longevity. 2018. pp. 63-98. DOI: 10.1007/978-3-030-03742-0_3

[39] Wu M, Porres JM, Cheng WH. Selenium, selenoproteins, and age-related disorders. In: Bioactive Food as Dietary Interventions for the Aging Population: Bioactive Foods in Chronic Disease States. 2012. p. 227. DOI: 10.1016/B978-0-12-397155-5.00019-2

[40] Lubos E, Loscalzo J, Handy DE. Glutathione peroxidase-1 in health and disease: From molecular mechanisms to therapeutic opportunities. Antioxidants & Redox Signaling. 2011;**15**(7):1957-1997. DOI: 10.1089/ars.2010.3586

[41] Arthur JR. The Glutathione Peroxidases. Cellular and Molecular Life Sciences CMLS. 2000;**57**:1825-1835. DOI: 10.1007/PL00000664

[42] Flohé L, Toppo S, Orian L. The glutathione peroxidase family: Discoveries and mechanism. Free Radical Biology & Medicine. 2022;**187**:113-122. DOI: 10.1016/j.freeradbiomed.2022.05.003

[43] Avissar N, Ornt DB, Yagil Y, et al. Human kidney proximal tubules are the main source of plasma glutathione peroxidase. American Journal of Physiology. Cell Physiology. 1994;**266**(2):35-32. DOI: 10.1152/ajpcell.1994.266.2.c367

[44] Schmutzler C, Mentrup B, Schomburg L, Hoang-Vu C, Herzog V, Köhrle J. Selenoproteins of the thyroid gland: Expression, localization and possible function of glutathione peroxidase 3. Biological Chemistry. 2007;**388**(10):1053-1059. DOI: 10.1515/BC.2007.122

[45] Chung SS, Kim M, Youn BS, et al. Glutathione peroxidase 3 mediates the antioxidant effect of peroxisome

proliferator-activated receptor γ in human skeletal muscle cells. Molecular and Cellular Biology. 2009;**29**(1):20-30. DOI: 10.1128/mcb.00544-08

[46] Imai H, Matsuoka M, Kumagai T, Sakamoto T, Koumura T. Lipid peroxidation-dependent cell death regulated by GPx4 and ferroptosis. Current Topics in Microbiology and Immunology. 2017;**403**. DOI: 10.1007/82_2016_508

[47] Kryukov GV, Castellano S, Novoselov SV, et al. Characterization of mammalian selenoproteomes. Science (1979). 2003;**300**(5624):1439-1443. DOI: 10.1126/science.1083516

[48] Zeng R, Farooq MU, Zhang G, et al. Dissecting the potential of Selenoproteins extracted from selenium-enriched Rice on physiological, biochemical and anti-ageing effects In vivo. Biological Trace Element Research. 2020;**196**(1):119-130. DOI: 10.1007/s12011-019-01896-z

[49] Lu J, Holmgren A. The thioredoxin antioxidant system. Free Radical Biology & Medicine. 2014;**66**:75-87. DOI: 10.1016/j.freeradbiomed.2013.07.036

[50] Papp LV, Lu J, Holmgren A, Khanna KK. From selenium to selenoproteins: Synthesis, identity, and their role in human health. Antioxidants & Redox Signaling. 2007;**9**(7):775-806. DOI: 10.1089/ars.2007.1528

[51] Holmgren A. Thioredoxin and glutaredoxin systems. Journal of Biological Chemistry. 1989;**264**(24):13963-13966

[52] Bianco AC. Minireview: Cracking the metabolic code for thyroid hormone signaling. Endocrinology. 2011;**152**(9):3306-3311. DOI: 10.1210/en.2011-1104

[53] St. Germain DL, Hernandez A, Schneider MJ, Galton VA. Insights into the role of deiodinases from studies of genetically modified animals. Thyroid. 2005;**15**(8). DOI: 10.1089/thy.2005.15.905

[54] Köhrle J, Jakob F, Contempré B, Dumont JE. Selenium, the thyroid, and the endocrine system. Endocrine Reviews. 2005;**26**(7). DOI: 10.1210/er.2001-0034

[55] Low SC, Harney JW, Berry MJ. Cloning and functional characterization of human selenophosphate synthetase, an essential component of selenoprotein synthesis. Journal of Biological Chemistry. 1995;**270**(37). DOI: 10.1074/jbc.270.37.21659

[56] Tamura T, Yamamoto S, Takahata M, et al. Selenophosphate synthetase genes from lung adenocarcinoma cells: Sps1 for recycling L-selenocysteine and Sps2 for selenite assimilation. Proceedings of the National Academy of Sciences of the United States of America. 2004;**101**(46). DOI: 10.1073/pnas.0406313101

[57] Stadtman ER. Protein oxidation and aging. Free Radical Research. 2006;**40**(12). DOI: 10.1080/10715760600918142

[58] Behne D, Kyriakopoulos A, Kalcklösch M, et al. Two new Selenoproteins found in the prostatic glandular epithelium and in the spermatid nuclei. Biomedical and Environmental Sciences. 1997;**10**(2-3)

[59] Kumaraswamy E, Malykh A, Korotkov KV, et al. Structure-expression relationships of the 15-kDa selenoprotein gene: Possible role of the protein in cancer etiology. Journal of Biological Chemistry. 2000;**275**(45). DOI: 10.1074/jbc.M004014200

[60] Labunskyy VM, Hatfield DL, Gladyshev VN. The Sep15 protein family:

Roles in disulfide bond formation and quality control in the endoplasmic reticulum. IUBMB Life. 2007;**59**(1). DOI: 10.1080/15216540601126694

[61] Ferguson AD, Labunskyy VM, Fomenko DE, et al. NMR structures of the selenoproteins Sep15 and SelM reveal redox activity of a new thioredoxin-like family. Journal of Biological Chemistry. 2006;**281**(6). DOI: 10.1074/jbc. M511386200

[62] Lu C, Qiu F, Zhou H, et al. Identification and characterization of selenoprotein K: An antioxidant in cardiomyocytes. FEBS Letters. 2006;**580**(22). DOI: 10.1016/j. febslet.2006.08.065

[63] Korotkov KV, Novoselov SV, Hatfield DL, Gladyshev VN. Mammalian Selenoprotein in which Selenocysteine (sec) incorporation is supported by a new form of sec insertion sequence element. Molecular and Cellular Biology. 2002;**22**(5). DOI: 10.1128/mcb.22.5.1402-1411.2002

[64] Hwang DY, Cho JS, Oh JH, et al. Differentially expressed genes in transgenic mice carrying human mutant presenilin-2 (N141I): Correlation of selenoprotein M with Alzheimer's disease. Neurochemical Research. 2005;**30**(8). DOI: 10.1007/ s11064-005-6787-6

[65] Castets P, Lescure A, Guicheney P, Allamand V. Selenoprotein N in skeletal muscle: From diseases to function. Journal of Molecular Medicine. 2012;**90**(10). DOI: 10.1007/ s00109-012-0896-x

[66] Castets P, Maugenre S, Gartioux C, et al. Selenoprotein N is dynamically expressed during mouse development and detected early in muscle precursors. BMC Developmental Biology. 2009;**9**(1). DOI: 10.1186/1471-213X-9-46

[67] Arbogast S, Ferreiro A. Selenoproteins and protection against oxidative stress: Selenoprotein N as a novel player at the crossroads of redox signaling and calcium homeostasis. Antioxidants & Redox Signaling. 2010;**12**(7). DOI: 10.1089/ars.2009.2890

[68] Moghadaszadeh B, Rider BE, Lawlor MW, et al. Selenoprotein N deficiency in mice is associated with abnormal lung development. FASEB Journal. 2013;**27**(4). DOI: 10.1096/ fj.12-212688

[69] Han SJ, Lee BC, Yim SH, Gladyshev VN, Lee SR. Characterization of mammalian selenoprotein O: A redox-active mitochondrial protein. PLoS One. 2014;**9**(4). DOI: 10.1371/journal. pone.0095518

[70] Sreelatha A, Yee SS, Lopez VA, et al. Protein AMPylation by an evolutionarily conserved Pseudokinase. Cell. 2018;**175**(3). DOI: 10.1016/j.cell.2018.08.046

[71] Burk RF, Hill KE. Selenoprotein P-expression, functions, and roles in mammals. Biochimica et Biophysica Acta - General Subjects. 2009;**1790**(11). DOI: 10.1016/j.bbagen.2009.03.026

[72] Turanov AA, Everley RA, Hybsier S, et al. Regulation of selenocysteine content of human selenoprotein p by dietary selenium and insertion of cysteine in place of selenocysteine. PLoS One. 2015;**10**(10). DOI: 10.1371/journal.pone.0140353

[73] Steinbrenner H, Alili L, Bilgic E, Sies H, Brenneisen P. Involvement of selenoprotein P in protection of human astrocytes from oxidative damage. Free Radical Biology & Medicine. 2006;**40**(9). DOI: 10.1016/j. freeradbiomed.2005.12.022

[74] Steinbrenner H, Bilgic E, Alili L, Sies H, Brenneisen P. Selenoprotein P protects

endothelial cells from oxidative damage by stimulation of glutathione peroxidase expression and activity. Free Radical Research. 2006;**40**(9). DOI: 10.1080/10715760600806248

[75] Atkinson JB, Hill KE, Burk RF. Centrilobular endothelial cell injury by diquat in the selenium-deficient rat liver. Laboratory Investigation. 2001;**81**(2). DOI: 10.1038/labinvest.3780227

[76] Traulsen H, Steinbrenner H, Buchczyk DP, Klotz LO, Sies H. Selenoprotein P protects low-density lipoprotein against oxidation. Free Radical Research. 2004;**38**(2). DOI: 10.1080/10715760320001634852

[77] Steinbrenner H, Speckmann B, Klotz LO. Selenoproteins: Antioxidant selenoenzymes and beyond. Archives of Biochemistry and Biophysics. 2016;**595**. DOI: 10.1016/j.abb.2015.06.024

[78] Ghelichkhani F, Gonzalez FA, Kapitonova MA, Rozovsky S. Selenoprotein S interacts with the replication and transcription complex of SARS-CoV-2 by binding nsp7. Journal of Molecular Biology. 2023;**435**(8):168008. DOI: 10.1016/j.jmb.2023.168008

[79] Pothion H, Jehan C, Tostivint H, et al. Selenoprotein T: An essential oxidoreductase serving as a Guardian of endoplasmic reticulum homeostasis. Antioxidants & Redox Signaling. 2020;**33**(17):1257-1275. DOI: 10.1089/ars.2019.7931

[80] Zhang X, Xiong W, Chen LL, Huang JQ, Lei XG. Selenoprotein V protects against endoplasmic reticulum stress and oxidative injury induced by pro-oxidants. Free Radical Biology & Medicine. 2020;**160**:670-679. DOI: 10.1016/j.freeradbiomed.2020.08.011

[81] Whanger PD. Selenoprotein W: A review. Cellular and Molecular Life Sciences. 2000;**57**(13-14). DOI: 10.1007/PL00000666

[82] Loflin J, Lopez N, Whanger PD, Kioussi C. Selenoprotein W during development and oxidative stress. Journal of Inorganic Biochemistry. 2006;**100**(10). DOI: 10.1016/j.jinorgbio.2006.05.018

[83] Duntas LH. Selenium and inflammation: Underlying anti-inflammatory mechanisms. Hormone and Metabolic Research. 2009;**41**(6). DOI: 10.1055/s-0029-1220724

[84] Tsuji PA, Carlson BA, Anderson CB, Seifried HE, Hatfield DL, Howard MT. Dietary selenium levels affect selenoprotein expression and support the interferon-γ and IL-6 immune response pathways in mice. Nutrients. 2015;7(8). DOI: 10.3390/nu7085297

[85] Bentley-Hewitt KL, Chen RKY, Lill RE, et al. Consumption of selenium-enriched broccoli increases cytokine production in human peripheral blood mononuclear cells stimulated ex vivo, a preliminary human intervention study. Molecular Nutrition & Food Research. 2014;**58**(12). DOI: 10.1002/mnfr.201400438

[86] Hoffmann FKW, Hashimoto AC, Shafer LA, Dow S, Berry MJ, Hoffmann PR. Dietary selenium modulates activation and differentiation of CD4 + T cells in mice through a mechanism involving cellular free thiols. Journal of Nutrition. 2010;**140**(6). DOI: 10.3945/jn.109.120725

[87] Ravaglia G, Forti P, Maioli F, et al. Effect of micronutrient status on natural killer cell immune function in healthy free-living subjects aged ≥90 y. American Journal of Clinical Nutrition. 2000;**71**(2):590-598. DOI: 10.1093/ajcn/71.2.590

[88] Mahmoodpoor A, Hamishehkar H, Shadvar K, Ostadi Z, Sanaie S, Saghaleini SH, et al. The effect of intravenous selenium on oxidative stress in critically ill patients with acute respiratory distress syndrome. Immunological Investigations. 2009;**48**(2):147-159. DOI: 10.1080/08820139.2018.1496098

[89] Razaghi A, Poorebrahim M, Sarhan D, Björnstedt M. Selenium stimulates the antitumour immunity: Insights to future research. European Journal of Cancer. 2021;**155**:256-267. DOI: 10.1016/j.ejca.2021.07.013

[90] Carlson BA, Yoo MH, Shrimali RK, et al. Role of selenium-containing proteins in T-cell and macrophage function. Proceedings of the Nutrition Society. 2010;**69**:300-310. DOI: 10.1017/S002966511000176X

[91] Marciel MP, Hoffmann PR. Molecular mechanisms by which Selenoprotein K regulates immunity and cancer. Biological Trace Element Research. 2019;**192**(1):60-68. DOI: 10.1007/s12011-019-01774-8

Section 2

Role of Selenium in Various Diseases

Chapter 3

Vascular System: Role of Selenium in Vascular Diseases

Muhammed Fatih Doğan

Abstract

The trace element selenium is crucial for cellular defense against oxidative stress and inflammatory reactions. Balanced selenium levels are important for the vascular system, whereas dysregulation can damage vascular reactivity. Reports have also supported the strong relationship between oxidative stress and vascular inflammation, which are induced by either the overproduction of reactive oxygen species (ROS) or the lack of antioxidant defense proteins. The damage of vascular smooth muscle and endothelium layer are frequently linked to vascular disorders such as hypertension, hypercholesterolemia, and atherosclerosis. Vascular diseases can result in life-threatening serious cardiovascular complications, such as blood clots, heart attack, and stroke. Selenium levels are crucial for preventing vascular damage; however, either low or extremely high amounts of selenium intake may contribute to the pathophysiology of vascular disorders. Selenoproteins are proteins such as glutathione peroxidase containing selenium in the form of the 21st amino acid, selenocysteine. Selenoproteins have the capacity to protect vascular smooth muscle and endothelium by lowering harmful ROS, which allows them to regulate normal vascular functions including vasoreactivity. The current chapter's goal was to carry out a thorough evaluation of the literature on the connection between selenium and vascular disorders.

Keywords: selenium, selenoproteins, vascular system, vascular disease, hypertension

1. Introduction

Selenium is a cofactor of enzymes that are responsible for antioxidant protection in the body. It is abundant in the environment at varying levels and plays an important role in the regulation of inflammatory processes in the body [1]. Adequate bioavailable levels of selenium in the organism are functionally important for many aspects of human biology, including the cardiovascular system, central nervous system, male reproductive biology, endocrine system, muscle function, and immunity [2]. Selenium is an essential component of selenoproteins, which play an important role in a variety of biological functions including antioxidant defense, thyroid hormone formation, DNA synthesis, fertility, and reproduction [3]. Many selenoproteins have been identified in the organism, including glutathione peroxidases (GPXs), thioredoxin reductase (TrxR), iodothyronine deiodinase, selenoprotein P, and selenoprotein W [4]. GPXs of the selenoprotein family are antioxidants that play an important role in oxidative stress and vascular tissue damage [5]. When the oxidative-antioxidant balance function is disrupted as a result of oxidative stress, several pathogenic processes can occur in vascular system. Oxidative stress and the formation of reactive oxygen species (ROS) contribute to the progression of tissue injury by activating the inflammatory response via the release of proinflammatory cytokines and the accumulation of inflammatory cells in tissues [6]. Endothelial dysfunction is important in the development of vascular diseases, and selenium reduces endothelial damage and prevents disruption of endothelial-dependent relaxation [7]. While a lack of selenium can lead to a variety of diseases, an excess of selenium can be toxic and result in the selenosis condition [8]. Low or excessive selenium consumption has been linked to different vascular diseases such as hypertension, hypercholesterolemia, and atherosclerosis [9]. The fundamental pharmacological, physiological, and pathophysiological properties of selenium in vascular disease are presented in this chapter.

2. The relationship between selenium and vascular diseases

Vascular smooth muscle (VSM) tone in the arterial vessels determines peripheral vascular resistance and blood pressure. Endothelial cells regulate VSM tone and subsequently blood flow by producing and releasing relaxants such as nitric oxide and contractile substances such as endothelin. The defective function of VSM and endothelial layer are commonly associated with impaired vascular responses [10]. The coexistence of dyslipidemia and oxidative stress is a major risk factor for the development of vascular diseases such as atherosclerosis and hypertension [11]. VSM and endothelial cells function properly and maintain an appropriate oxidant/antioxidant balance when selenium and selenoproteins are present in the proper amounts [12]. A sufficient concentration of selenium-dependent GPXs is required to maintain an active endogenous antioxidant system, which prevents vascular diseases caused by hypertension, hypercholesterolemia, and atherosclerosis [13]. Overall, the GPX family is one of the best-studied selenoprotein families in cardiovascular biology. There are five different types of GPX isoforms, with GPx-3 being the only one found in the extracellular space [14]. GPx-3 deficiency causes a prothrombotic state and vascular dysfunction, which promotes platelet-dependent arterial thrombosis [9]. The link between low selenium intake and cardiovascular pathologies is due to increased oxidative stress and its consequences in the development of non-infectious vascular diseases [15]. Decreased amount of selenium in the body is associated with an increase in adhesion molecules and a decrease in the expression of selenoproteins despite endothelial cell integrity and function [16]. A potentially harmful relationship was discovered between high selenium levels and carotid wall thickening, despite a long-term vascular protective effect between arterial stiffness and blood pressure in people with normal selenium levels [17]. All of this points to a significant relationship between selenium and the vascular system. **Table 1** summarizes studies demonstrating the effect of selenium on vascular diseases.

2.1 Hypertension

Because of its high prevalence and associated risks of cardiovascular and kidney disease, hypertension is a major public health issue worldwide [30]. Endothelial dysfunction, inflammation, hypertrophy, apoptosis, cell migration, fibrosis, and angiogenesis have all been linked to vascular remodeling in hypertension [31].

Disease	Species	Method	Conclusion	References
Hypertension	Human	Selenium deficiency	Higher risk in pregnancy-induced hypertension	[18]
Hypertension	Human	Higher selenium levels	Lower risk in ischemic stroke	[19]
Hypertension	Human	Higher selenium levels	Selenium protects vascular function	[20]
Hypertension	Human	Higher selenium levels	Higher prevalence of hypertension	[21]
Hypertension	Rat	Selenium deficiency	Increase in AT1 receptors and blood pressure	[22]
Hypertension	Rat	Higher selenium levels	Increase in systolic blood pressure	[23]
Dyslipidemia	Human	Selenium deficiency	Low HDL levels, high LDL levels	[24]
Hyperlipidemia	Human	Selenium deficiency	Selenium may prevent hyperlipidemia	[25]
Dyslipidemia and atherosclerosis	Rabbit	0.5% dietary cholesterol-induced dyslipidemic rabbits	Co-supplementation of vitamin K2 and selenium improved metabolic profile and atherosclerosis	[26]
Hyperlipidemia	Mice	Selenium nanoparticles	Selenium reduces hyperlipidemia and vascular injury	[27]
Hyperlipidemia	Mice	High fat diet-induced dyslipidemia	Selenium-rich *Cordyceps militaris* polysaccharides could prevent hyperlipidemia	[28]
Endothelial dysfunction	Rat	Homocysteine-induced endothelial dysfunction and apoptosis	Selenium protects against homocysteine-induced dysfunction and apoptosis of endothelial cells.	[7]
Endothelial dysfunction	Rat	Streptozotocin-induced diabetic aorta	Selenium improved vascular responses and endothelial dysfunction	[29]

Table 1.
Selenium research in vascular disease.

Ozturk et al. reported that selenium reduced the disruption of endothelium-dependent vasorelaxation in the diabetic aorta and improved vascular responses and endothelial dysfunction in diabetes by regulating antioxidant enzymes and nitric oxide release [29]. Many studies have been conducted to investigate the relationship between hypertension and low and high dietary selenium intake. Selenium deficiency in rats caused an increase in H2O2 production by decreasing GPx1 expression and increased renal angiotensin II type 1 receptor expression by increasing NF-κB activity, resulting in sodium retention and an increase in blood pressure [22]. It has been reported that men with antioxidant selenium deficiency (selenium concentration lower than 20 μg/l) have higher blood pressure and a higher risk of developing hypertension [32]. Obese elderly people may require more antioxidants, particularly selenium, to counteract the increased oxidative stress that leads to vascular oxidative dysfunction [33]. Increased plasma selenium levels were found to be significantly associated with a lower risk of first stroke and ischemic stroke in hypertensive adults [19]. Selenium has been shown to lower the incidence of mercury-related hypertension and protect vascular function among Inuit in Canada [20]. Lower serum selenium levels in early healthy pregnancy were linked to an increased risk of pregnancy-induced hypertension and served as a risk marker for this potentially dangerous disease [18]. High selenium intake appeared to be a blood pressure protective factor, particularly in people living in low selenium areas [34]. On the contrary, some studies have found that a high selenium intake is associated with vascular system damage. Increasing selenium levels above the recommended daily intake is not beneficial for vascular health and may even cause hypertension, hyperlipidemia, and diabetes [35]. According to *Laclaustra et al.*, there is a strong correlation between elevated serum selenium levels and a high prevalence of hypertension in the US population [21]. Similarly, long-term selenium supplementation (2 and 6 mg/L) resulted in a significant increase in systolic blood pressure in rats after 42 days [23]. The cause of hypertension caused by high selenium intake may be related to endothelial dysfunction via a mechanism involving cell death mediated by ROS production induced by endoplasmic reticulum stress [36]. While a high selenium intake is generally beneficial through an antioxidant mechanism, it may also be a factor in the development of hypertension.

2.2 Hypercholesterolemia and atherosclerosis

Hyperlipidemia, caused by hypercholesterolemia and/or hypertriglyceridemia, is a critical condition that plays a significant role in the pathogenesis of atherosclerosis [37]. Apoptotic VSM cells, which are found in advanced atherosclerosis, cause plaque instability and rupture, which results in thrombosis and the clinical symptoms of a heart attack or stroke [38]. Selenoproteins can destroy cholesterol that has accumulated in the vascular lumen. Inadequate plasma selenium levels can lead to vascular disease by lowering selenoprotein levels [39]. Optimal Se uptake prevents atherosclerosis by reducing oxidative stress, inflammation, endothelial dysfunction, vascular cell apoptosis, and vascular calcification [40]. Selenium supplementation increases GPX1, GPX4, and TRXR1 expression and activity in vascular endothelial or smooth muscle cells. As a result, it prevents oxidative stress, cell damage, and apoptosis caused by oxidized low-density lipoprotein (LDL), a cytotoxic hydroxylated cholesterol derivative found in human blood, cells, tissues, and atherosclerotic plaques [41]. The selenoenzyme GPX uses GSH as an electron donor to neutralize hydroperoxide and protects against arsenic-induced atherosclerosis in a mouse model [42]. In healthy young subjects, a negative relationship was found between serum triglycerides and sialic acid, an inflammation marker, and dietary selenium intake [43]. Experimental studies have shown that selenium is used in combination with other substances and improves hyperlipidemia more strongly. It was reported that concomitant administration of vitamin K2 and selenium improved metabolic function, markers of cardiovascular health, and atherosclerosis in dyslipidemic rabbits [26]. *Yu et al.* also found that consuming a high dose of selenium-rich *Cordyceps militaris* polysaccharides could prevent high fat diet-induced dyslipidemia and dysbiosis of the gut microbiota, and that it could be used as a functional food [28]. Vascular dysfunction occurs in patients with a high selenium deficiency, and there is a positive correlation between HDL and selenium in dyslipidemic patients [24]. Selenium levels tend to decrease with age, and high selenium status may be beneficial in preventing hyperlipidemia in young adult females [25]. Selenium nanoparticles could significantly reduce hyperlipidemia and vascular injury in apolipoprotein E deficient mice, possibly by regulating cholesterol metabolism and reducing oxidative stress via antioxidant selenoenzymes/selenoproteins, and could be a potential candidate for atherosclerosis prevention [27].

3. Conclusions

These findings suggest that adequate selenium intake may contribute to preventing the development of hypertension, hyperlipidemia, and atherosclerosis by reducing oxidative stress and inflammation associated with vascular diseases. Furthermore, the findings emphasize the importance of consuming or supplementing with an adequate amount of selenium to optimize vascular system function. Selenium appears to have both a protective and a therapeutic role in the vascular dysfunction, and more research on the effect of selenium on the vascular system is required. As a result of the recent studies, it is understood that people living in low selenium-containing regions should be protected from vascular damage by taking selenium supplements. Selenium, which regulates blood pressure and reduces atherosclerosis caused by hyperlipidemia, could be used as a potential pharmacological agent in the prevention of vascular diseases in the near future.

Author details

Muhammed Fatih Doğan
Faculty of Medicine, Department of Pharmacology, Pamukkale University, Denizli, Turkey

*Address all correspondence to: mfatihdogan89@gmail.com

References

[1] Kieliszek M. Selenium. Advances in Food and Nutrition Research. 2021;**96**:417-429

[2] Avery JC, Hoffmann PR. Selenium, selenoproteins, and immunity. Nutrients. 2018;**10**(9):1203

[3] Mehdi Y, Hornick JL, Istasse L, Dufrasne I. Selenium in the environment, metabolism and involvement in body functions. Molecules. 2013;**18**(3):3292-3311

[4] Sur Ü, Erkekoğlu P, Koçer-Gümüşel B. Selenoproteinler ve Hashimoto Tiroiditi. Journal of Pharmaceutical Sciences. 2020;**45**(1):45-64

[5] Brigelius-Flohé R, Maiorino M. Glutathione peroxidases. Biochimica et Biophysica Acta - General Subjects. 2013;**1830**(5):3289-3303

[6] Doğan MF, Kaya K, Demirel HH, Şahin Y, Çiftçi O. The effect of vitamin C supplementation on favipiravir-induced oxidative stress and proinflammatory damage in livers and kidneys of rats and proinflammatory damage in livers and kidneys of rats. Immunopharmacology and Immunotoxicology. 2023:1-6

[7] Ren H et al. Selenium inhibits homocysteine-induced endothelial dysfunction and apoptosis via activation of AKT. Cellular Physiology and Biochemistry. 2016;**38**(3):871-882

[8] Hariharan S, Dharmaraj S. Selenium and selenoproteins: it's role in regulation of inflammation. Inflammopharmacology. 2020;**28**(3):667-695

[9] Shimada BK, Alfulaij N, Seale LA. The impact of selenium deficiency on cardiovascular function. International Journal of Molecular Sciences. 2021;**22**(19):10713

[10] Dogan MF, Yildiz O, Arslan SO, Ulusoy KG. Potassium channels in vascular smooth muscle: A pathophysiological and pharmacological perspective. Fundamental & Clinical Pharmacology. 2019;**33**(5):504-523

[11] Holvoet P. Relations between metabolic syndrome, oxidative stress and inflammation and cardiovascular disease. Verhandelingen - Koninklijke Academie voor Geneeskunde van België. 2008;**70**(3):193-219

[12] Gać P et al. The importance of selenium and zinc deficiency in cardiovascular disorders. Environmental Toxicology and Pharmacology. 2021;**82**:103553

[13] Jenkins DJA et al. Selenium, antioxidants, cardiovascular disease, and all-cause mortality: A systematic review and meta-analysis of randomized controlled trials. The American Journal of Clinical Nutrition. 2020;**112**(6):1642-1652

[14] Jin RC et al. Glutathione peroxidase-3 deficiency promotes platelet-dependent thrombosis in vivo. Circulation. 2011;**123**(18):1963-1973

[15] Benstoem C et al. Selenium and its supplementation in cardiovascular disease—What do we know? Nutrients. 2015;**7**(5):3094-3118

[16] Lopes Junior E, Leite HP, Konstantyner T. Selenium and selenoproteins: From endothelial cytoprotection to clinical outcomes. Translational Research. 2019;**208**:85-104

[17] Swart R, Schutte AE, van Rooyen JM, Mels CMC. Selenium and large artery structure and function: A 10-year

prospective study. European Journal of Nutrition. 2019;**58**(8):3313-3323

[18] Lewandowska M, Sajdak S, Lubiński J. Serum selenium level in early healthy pregnancy as a risk marker of pregnancy induced hypertension. Nutrients. 2019;**11**(5):1028

[19] Wang Z et al. Plasma selenium and the risk of first stroke in adults with hypertension: A secondary analysis of the China stroke primary prevention trial. The American Journal of Clinical Nutrition. 2022;**115**(1):222-231

[20] Hu XF, Eccles KM, Chan HM. High selenium exposure lowers the odds ratios for hypertension, stroke, and myocardial infarction associated with mercury exposure among Inuit in Canada. Environment International. 2017;**102**:200-206

[21] Laclaustra M, Navas-Acien A, Stranges S, Ordovas JM, Guallar E. Serum selenium concentrations and hypertension in the US population. Circulation. Cardiovascular Quality and Outcomes. 2009;**2**(4):369-376

[22] Lei L et al. Selenium deficiency causes hypertension by increasing renal AT1 receptor expression via GPx1/H2O2/NF-κB pathway. Free Radical Biology & Medicine. 2023;**200**:59-72

[23] Grotto D, Carneiro MFH, de Castro MM, Garcia SC, Junior FB. Long-term excessive selenium supplementation induces hypertension in rats. Biological Trace Element Research. 2018;**182**(1):70-77

[24] Arnaud J, Akbaraly TN, Hininger-Favier I, Berr C, Roussel AM. Fibrates but not statins increase plasma selenium in dyslipidemic aged patients - the EVA study. Journal of Trace Elements in Medicine and Biology. 2009;**23**(1):21-28

[25] Lee O, Moon J, Chung Y. The relationship between serum selenium levels and lipid profiles in adult women. Journal of Nutritional Science and Vitaminology (Tokyo). 2003;**49**(6):397-404

[26] Atteia HH. Co-supplementation of vitamin K2 and selenium synergistically improves metabolic status and reduces cardiovascular risk markers in Dyslipidemic rabbits. Biological Trace Element Research. 2023:1-15

[27] Guo L, Xiao J, Liu H, Liu H. Selenium nanoparticles alleviate hyperlipidemia and vascular injury in ApoE-deficient mice by regulating cholesterol metabolism and reducing oxidative stress. Metallomics. 2020;**12**(2):204-217

[28] Yu M et al. Anti-hyperlipidemia and gut microbiota community regulation effects of selenium-rich cordyceps militaris polysaccharides on the high-fat diet-fed mice model. Food. 2021;**10**(10):2252

[29] Ozturk Z, Gurpinar T, Vural K, Boyacioglu S, Korkmaz M, Var A. Effects of selenium on endothelial dysfunction and metabolic profile in low dose streptozotocin induced diabetic rats fed a high fat diet. Biotechnic & Histochemistry. 2015;**90**(7):506-515

[30] Kearney PM, Whelton M, Reynolds K, Muntner P, Whelton PK, He J. Global burden of hypertension: Analysis of worldwide data. Lancet. 2005;**365**(9455):217-223

[31] Sinha N, Dabla P. Oxidative stress and antioxidants in hypertension–A current review. Current Hypertension Reviews. 2015;**11**(2):132-142

[32] Nawrot TS et al. Blood pressure and blood selenium: A cross-sectional and longitudinal population study. European Heart Journal. 2007;**28**(5):628-633

[33] Arnaud J, Akbaraly NT, Hininger I, Roussel AM, Berr C. Factors associated with longitudinal plasma selenium decline in the elderly: The EVA study. The Journal of Nutritional Biochemistry. 2007;**18**(7):482-487

[34] Xie C et al. Regional difference in the association between the trajectory of selenium intake and hypertension: A 20-year cohort study. Nutrients. 2021;**13**(5):1501

[35] Kuruppu D, Hendrie HC, Yang L, Gao S. Selenium levels and hypertension: A systematic review of the literature. Public Health Nutrition. 2014;**17**(6):1342-1352

[36] Zachariah M, Maamoun H, Milano L, Rayman MP, Meira LB, Agouni A. Endoplasmic reticulum stress and oxidative stress drive endothelial dysfunction induced by high selenium. Journal of Cellular Physiology. 2021;**236**(6):4348-4359

[37] Tietge UJF. Hyperlipidemia and cardiovascular disease: Inflammation, dyslipidemia, and atherosclerosis. Current Opinion in Lipidology. 2014;**25**(1):94-95

[38] Kockx MM, Herman AG. Apoptosis in atherosclerosis: Beneficial or detrimental? Cardiovascular Research. 2000;**45**(3):736-746

[39] Kielczykowska M, Kocot J, Pazdzior M, Musik I. Selenium - a fascinating antioxidant of protective properties. Advances in Clinical and Experimental Medicine. 2018;**27**(2):245-255

[40] Liu H, Xu H, Huang K. Selenium in the prevention of atherosclerosis and its underlying mechanisms. Metallomics. 2017;**9**(1):21-37

[41] Bellinger FP, Raman AV, Reeves MA, Berry MJ. Regulation and function of selenoproteins in human disease. Biochemical Journal. 2009;**422**(1):11-22

[42] Krohn RM et al. High-selenium lentil diet protects against arsenic-induced atherosclerosis in a mouse model. The Journal of Nutritional Biochemistry. 2016;**27**:9-15

[43] Zulet MÁ, Puchau B, Hermsdorff HHM, Navarro C, Martínez JA. Dietary selenium intake is negatively associated with serum sialic acid and metabolic syndrome features in healthy young adults. Nutrition Research. 2009;**29**(1):41-48

Chapter 4

Efficacy of Selenium for Controlling Infectious Diseases

Poonam Gopika Vinayamohan, Divya Joseph, Leya Susan Viju and Kumar Venkitanarayanan

Abstract

Selenium, an essential micronutrient for both animals and humans, has been documented to possess antimicrobial properties against a wide range of pathogenic microorganisms. One of the primary mechanisms by which selenium exerts its antimicrobial activity is through the generation of reactive oxygen species that can damage microbial cells. Besides its direct antimicrobial effects, selenium can enhance the immune response to infections, making it a potential tool in the prevention and treatment of infectious diseases. Given the growing threat of antibiotic resistance and the need for alternative therapeutic options, the antibacterial properties of selenium are of interest to the scientific community. This book chapter will summarize the current state of knowledge on the antibacterial properties of selenium, and its potential clinical applications as a therapeutic agent against infectious diseases. Further, the chapter explores the limitations and challenges associated with the use of selenium as an antibacterial agent.

Keywords: selenium, nanoparticles, immune response, antimicrobial effect, human health

1. Introduction

Selenium, a trace element discovered in 1817 by the Swedish chemist Jöns Jacob Berzelius, has since been demonstrated to be an indispensable micronutrient for human health. Although initially recognized for its practical value in preventing nutritional myopathies and vascular disorders in livestock, subsequent research revealed the numerous ways in which selenium contributes to overall human health and well-being.

Selenium has emerged as an essential component of several selenoproteins that play a crucial role in various physiological processes in humans. These processes include antioxidant defense, immune function, thyroid hormone metabolism, and redox homeostasis. The importance of selenium in human health became apparent when researchers discovered its role in glutathione peroxidase (GPx) in 1973, as well as its ability to prevent liver necrosis in vitamin E-deficient rats. This enzyme, which contains selenium as an integral part of its structure, is a potent antioxidant that neutralizes harmful reactive oxygen species (ROS) and protects cells from oxidative damage.

Selenium's role in human health received increased attention with the observation that selenium deficiency could lead to serious diseases such as Keshan disease, an endemic cardiomyopathy affecting people in selenium-deficient regions of China [1]. This discovery prompted further investigation into the geographical distribution of selenium intake and its impact on public health. Subsequent research has established that selenium deficiency is associated with a higher risk of certain cancers, impaired immune function, and cognitive decline. On the other hand, selenium toxicity, although rare, can occur when excessive amounts of the element are consumed, leading to conditions such as selenosis, which is characterized by symptoms such as hair loss, brittle nails, and gastrointestinal disturbances [2].

In recent years, the antimicrobial properties of selenium and its potential applications in combating pathogens of public health significance have become an area of growing interest. Recent advancements in nanotechnology have led to the development of selenium nanoparticles (SeNPs), which exhibit enhanced antimicrobial properties due to their increased surface area and unique physiochemical properties. SeNPs have been shown to exert direct antimicrobial effects, disrupt biofilms, and to improve host immune responses, making them a potential therapeutic agent against many pathogens.

Despite the growing body of evidence supporting selenium's antimicrobial properties, our understanding of its multifaceted functions in the human body remains incomplete. In this chapter, we will delve into the intricate mechanisms through which selenium exerts its immunomodulatory, antibacterial, and antiviral effects, and explore the potential applications of selenium in medicine and disease prevention. By providing a comprehensive understanding of the potential benefits of selenium in the context of human health and disease prevention, this chapter will shed light on its pivotal role in combating pathogens of public health significance.

2. Enhancement of immune response and combating pathogens with selenium and selenium nanoparticles

Selenium is renowned for its capacity to enhance immune responses against infections through multiple mechanisms. It can increase the number of T cells, improve the proliferative responses of lymphocytes to mitogens, stimulate the secretion of the cytokine IL-2, and enhance the activity of natural killer (NK) cells. These combined effects contribute to the strengthening of immune defences against various pathogens [3]. Selenium's ability to boost the immune system and reduce inflammation can be mainly attributed to its antioxidant properties where its primary role is to regulate the function of GPx. Gpx in turn, decreases the levels of hydrogen peroxide and phospholipid hydroperoxides, preventing the generation of free radicals and ROS [4]. It also decreases hydroperoxide intermediates in the metabolic pathway of arachidonic acid, consequently reducing the production of inflammatory prostaglandins and leukotrienes [5].

The main mechanism of action for selenium involves its interaction with selenoproteins, which include antioxidant enzymes like GPxs and thioredoxin reductases (TrxRs). Selenoproteins are composed of the amino acid selenocysteine (Sec), which is integrated into the protein structure during translation. This occurs after the conversion of O-phosphoseryl-transfer RNA (O-phosphoseryl-tRNA) [Ser]Sec into selenocysteyl tRNA[Ser]Sec [6]. Selenium deficiency as well as small changes in the expression and genetic variations of certain selenoproteins have been linked to cancer and immune dysfunction. Among the 25 genes encoding human selenoproteins, immune cells express most of them pointing towards its immune potential.

Notably, the GPx isoenzymes GPx1 and GPx4 exhibit the highest expression levels in both T lymphocytes and macrophages [7]. Studies have demonstrated that selenite supplementation can enhance the production of 15-deoxy-D(12,14)-prostaglandin J2, an anti-inflammatory compound derived from arachidonic acid, by upregulating prostaglandin D2 synthase. Additionally, selenite has been found to reduce the production of the proinflammatory prostaglandin E2 (PGE2) in murine macrophages [8]. Selenium supplementation in patients with low selenium status activated the proinflammatory cellular (Th1-type) immune response against pathogens, while preventing excessive immune system activation and tissue damage by favoring macrophage differentiation to the more anti-inflammatory M2 phenotype [9].

Research has indicated that the addition of selenium to poultry diets can result in elevated expression of interferon and ISG (interferon-stimulated genes) in lymphoid tissue cells playing a crucial role in enhancing the antiviral responses of these cells [5]. Additionally, selenium enhances the activity of various immune cells such as neutrophils, macrophages, NK cells, and T lymphocytes. It also promotes the production of antibodies and regulates the production of cytokines, including an increase in IL-2 and a reduction in TNF and IL-8. Moreover, selenium has preventive effects against inflammatory diseases by reducing the activation of Nuclear Factor kappa B (NF-κB) and the production of pro-inflammatory cytokines. Selenium exhibits cytotoxic effects and has the potential to induce apoptosis in tumor cells. Selenium also offers protection against UV radiation, reduces viral virulence, and contributes to the prevention of atherosclerosis and cardiovascular diseases [9].

Selenium nanoparticles (SeNPs) have also shown the capability to modulate autophagy in different cancer cells, a process commonly associated with the induction of cancer cell death or apoptosis. SeNPs lead to the formation of autophagosomes and enhance autophagy by regulating specific proteins involved in autophagy, such as Beclin-1, LC3-II, and p62 [10]. Importantly, autophagy plays a role in regulating immune functions that can impact the infection and survival of pathogens within host cells. Moreover, selenium nanoparticles have demonstrated significant immunomodulatory effects by influencing various immune cells and modulating essential signalling pathways associated with the immune response. With the emergence of chimeric antigen receptor T-cell (CAR-T) therapy, immunotherapy has become a promising new treatment for malignant tumors [11].

Studies have demonstrated that the inclusion of dietary chitosan-selenium nanoparticles (CTS-Se NPs) can improve the immune response and disease resistance in zebrafish when exposed to the bacterium *Aeromonas hydrophila* [12]. Following treatment with CTS-Se NPs, zebrafish splenocytes exhibited higher proliferation when stimulated with lipopolysaccharide (LPS) and concanavalin A (ConA). The immune response of splenocytes against ConA was found to be associated with the up-regulation in IL-2 and IL-12 production. Moreover, SeNPs can promote host antibacterial immunity by inducing host cell apoptosis, autophagy, and M1 anti-bacterial polarization, which significantly enhances the intracellular *Mycobacterium tuberculosis* killing efficiency [13].

3. Antibacterial activity of selenium and selenium nanoparticles

Selenium has recently gained attention for its potential antibacterial properties. Research has demonstrated its ability to interfere with the growth and metabolism of various bacterial species, making it a promising candidate for the prevention and treatment of bacterial infections. Selenium has been shown to inhibit the growth of several pathogenic bacteria, including *Staphylococcus aureus*, *Escherichia coli*, and *Helicobacter pylori*. Furthermore, selenium can enhance the antibacterial effects of conventional antibiotics, potentially reducing antibiotic resistance. This section delves into the mechanisms underlying selenium's antibacterial properties and its prospective applications in the prevention and treatment of bacterial infections.

In both eukaryotes and prokaryotes, selenium plays essential roles in diverse biological processes, including redox homeostasis, thyroid hormone metabolism, and immune function. Prokaryotes express a wide range of selenoproteins, with approximately 20% of sequenced prokaryotic genomes encoding at least one trait for selenium utilization. These selenoproteins participate in multiple selenium-dependent enzymes (such as formate dehydrogenase in *Methanococcus jannaschii* and glycine reductase in *Clostridioides difficile*) and may confer increased fitness to prokaryotes in the presence of selenium, similar to the benefits observed in humans and other mammals [14].

This intricate interplay between host and pathogen during infection poses a challenge for the mammalian host, as both parties compete for the limited selenium resources. Despite its importance, limited information is available regarding the role of selenium in bacterial physiology, virulence, and overall pathogenesis. The literature documenting the antimicrobial activity of selenium toward various pathogenic microorganisms is summarized below.

3.1 *Staphylococcus aureus*

Staphylococcus aureus is an opportunistic Gram-positive bacterium that can cause illnesses ranging from mild skin infections to more severe illnesses such as necrotizing pneumonia and bacteremia. Besides this, there is an increasing concern for antibiotic resistance among *S. aureus* including methicillin-resistant strains. As a result, there is a growing interest in exploring selenium as a potential therapeutic agent for controlling *S. aureus* infections [15].

The immune system's response to *S. aureus* infection involves the activation of NF-κB and mitogen-activated protein kinase (MAPK) signaling pathways, which play central roles in inflammation and the production of pro-inflammatory cytokines, including TNF-a, IL-1B, and IL-6 [16]. *S. aureus* has developed various strategies to evade the host's immune response, such as producing virulence factors to resist the mitochondrial agents generated by phagocytosis and competing with inducible nitric oxide synthase (iNOS) for the shared substrate arginine. Selenium, as an antioxidant and a vital component for optimal immune cell functioning, may aid in the host response to *S. aureus* infection.

Selenium-supplemented macrophages have been shown to produce reduced amounts of nitric oxide (NO) while increasing ROS production, particularly hydrogen peroxide. This supplementation also decreases bacterial arginase activity, limiting the bacterium's tolerance to oxidative stress. Furthermore, selenium enhances phagocytosis and increases the bactericidal capacity in a dose-dependent manner [15]. In the context of *S. aureus* infection, selenium supplementation has been found to decrease inflammatory cytokine gene expression and protein levels, such as TNF-a, IL-1b, and IL-6. Selenium inhibits the activation of both NF-κb and MAPK signaling pathways by suppressing the phosphorylation of IkBa, p65, Erk, jnk, and p38, thereby attenuating the overall inflammatory response [16].

Selenium has also been demonstrated to possess an immunoregulatory function on inflammation in mammary epithelial cells and glandular tissue during *S. aureus*-induced mastitis [17] and selenium supplementation was shown to decrease mastitis incidence in dairy cattle [18]. Selenium deficiency results in increased pro-inflammatory cytokine levels, while supplementation promotes anti-inflammatory cytokine expression and inhibits NF-κB activation [19]. Additionally, selenium inhibits *S. aureus* infection of the uterus and reduces the activation of toll-like receptor-2 (TLR-2) inflammatory signaling, decreasing caspase activity [20].

S. aureus is known to produce biofilms, which contribute to antibiotic resistance and chronic infections. The use of selenium nanoparticles (SeNPs) has shown promise in addressing this challenge. SeNPs have demonstrated significant inhibitory effects on *S. aureus* growth during the early stages of infection, potentially preventing biofilm formation [21]. Furthermore, SeNPs exhibit both anti-adherence and anti-microcolony formation properties against *S. aureus* biofilms indicating their potential to disrupt biofilm formation [22].

A practical application of SeNPs has been observed in coating titanium implants. These coatings have demonstrated potent antimicrobial activity against drug-resistant strains, such as methicillin-resistant *S. aureus* (MRSA) and methicillin-resistant *Staphylococcus epidermidis*. The SeNP-coated implants effectively inhibited biofilm formation and reduced bacterial viability [21]. This suggests the potential use of selenium nanoparticle coatings as an effective anti-infective barrier for orthopedic medical devices, offering a novel approach to combating biofilm-associated infections.

Diabetic foot wounds, which are often infected by antibiotic-resistant bacteria such as MRSA, require alternative antimicrobial drugs. A hybrid nanostructure comprising selenium, chitosan, and mupirocin has demonstrated significant antimicrobial activity against MRSA. This system played a crucial role in wound healing by reducing the minimum inhibitory concentrations (MIC) of mupirocin, and promoting wound contraction, angiogenesis, fibroblastosis, collagen production, and growth of hair follicle and epidermis [23].

Selenium holds promise as a therapeutic agent for controlling *S. aureus* infection, with research highlighting its potential in enhancing the immune response, preventing biofilm formation, and promoting wound healing. Additional studies are needed to ascertain the ideal dosage and explore its applications in clinical settings.

3.2 *Escherichia coli*

Escherichia coli is a Gram-negative bacterium that typically resides in the lower intestinal tract of humans and animals. Though the majority of *E. coli* are harmless, some can cause severe infections, such as gastrointestinal illness, urinary tract infections, and meningitis. The emergence of antibiotic-resistant strains of *E. coli* has led to a growing need for alternative treatments.

Selenium deficiency, especially in conjunction with vitamin E deficiency, has been found to exacerbate the pathology of gastrointestinal tract diseases caused by pathogenic *E. coli* such as those caused by enteropathogenic *E. coli* (EPEC) [24]. Deficiency in these nutrients leads to heightened oxidative stress, which in turn causes increased pro-inflammatory signaling and tissue damage. On the other hand, selenium-enriched probiotics have demonstrated protective effects against pathogenic *E. coli* in the gut, enhancing antioxidant performance, inhibiting pathogenic bacterial colonization, and bolstering immunity [25].

Selenium-enriched probiotics have been found to outperform sodium selenite in raising serum selenium levels, most likely due to the improved absorption of organic selenium compounds over inorganic ones [14]. These probiotics adhere to the intestine, effectively preventing pathogenic bacteria such as *E. coli* from interacting with potential binding sites. This emphasizes the capacity of selenium-enriched probiotics to support gut health by improving antioxidant performance, preventing pathogenic bacterial colonization, enhancing immunity, and reducing enteric illnesses.

Selenium supplementation has been reported to aid in the resolution of chronic bacterial prostatitis (CBP) caused by *E. coli*, especially when used in conjunction with antibiotics [26]. The current primary treatment against CBP involves the use of antibiotics, which necessitate small molecular weight and fat-soluble properties to facilitate diffusion across the prostate epithelial membrane. Combining selenium with the antibiotic ciprofloxacin resulted in a significant reduction of *E. coli* in the CBP model and a considerable decrease in inflammatory cell infiltration within the prostate tissue.

Selenium has also exhibited inhibitory effects on biofilm formation in uropathogenic *E. coli* (UPEC), which is responsible for 80% of urinary tract infections. Selenium reduces exopolysaccharide synthesis and downregulates biofilm-associated genes (*fim*A, *fim*H, *pap*G, *foc*A, *sfa*S) [27]. Moreover, it has proven effective in deactivating pre-established UPEC biofilms on urinary catheters.

In the context of enterohemorrhagic *E. coli* O157:H7, a foodborne pathogen, selenium has been shown to inhibit biofilm formation by reducing attachment, decreasing EPS production, and downregulating genes involved in biofilm production [28]. Additionally, selenium supplementation lowered extracellular and intracellular verotoxin levels, downregulated verotoxin genes, and reduced Gb3 receptor synthesis (receptor for verotoxin) in lymphoma cells by downregulating the LacCer synthase gene involved in Gb3 synthesis [29].

Although sodium selenite does not directly exhibit antibacterial properties against *E. coli* and other bacteria (*Bacillus subtilis*, *Bacillus mycoides*, and *Pseudomonas spp*.), it has been found to enhance the inhibitory effects of ampicillin and streptomycin on these bacterial growth [30]. This suggests that selenium supplementation may function as an adjuvant, complementing conventional antibiotic therapy in the treatment of *E. coli* infections.

3.3 *Helicobacter pylori*

Helicobacter pylori is a Gram-negative, microaerophilic, helix-shaped bacterium that colonizes the gastric mucous layer or adheres to the epithelial lining of the stomach [31]. Present in approximately 50% of the human population worldwide, *H. pylori* is responsible for causing 90% of duodenal ulcers and 80% of gastric ulcers [9], with infected individuals facing an increased risk of developing gastric cancer and mucosal-associated-lymphoid type lymphoma [31].

Currently, the treatment for *H. pylori* infection in humans involves a combination of proton pump inhibitors, amoxicillin, and clarithromycin [31]. However, *H. pylori* has shown to develop resistance to clarithromycin, leading to decreased eradication rates.

During *H. pylori* infection, micronutrient homeostasis, including that of selenium, is frequently disrupted, with equilibrium typically restored upon successful eradication of the pathogen [32]. Interestingly, whole plasma selenium level remains consistent between patients with or without *H. pylori* induced inflammation, and antral mucosa of individuals with *H. Pylori*-associated gastritis exhibits higher levels of selenium [33–35]. Moreover, increased inflammation scores of the antral mucosa correlate with elevated tissue selenium concentrations [33].

This increase in selenium concentration at the infected mucosa may be a protective response, where selenium acts as an antioxidant to prevent further damage caused by ROS or mediated the resolution of inflammation. This is supported by the decrease in gastric tissue selenium observed in patients after successful eradication of *H. pylori* [33]. A combination of antioxidants, including vitamins A, C, and E, and selenium, has been shown to protect against *H. pylori* infection and reduce gastritis severity in guinea pigs, highlighting the potential benefits of dietary antioxidant supplementation in the prevention and management of *H. pylori*-associated diseases [36].

It is essential to note that selenium deficiency has been identified as a risk factor for the conversion of precancerous gastric lesions into carcinomas [33]. The decline in selenium may be due to long-lasting mucosal inflammation, which results in an altered gastric microenvironment leading to gastric carcinogenesis. These findings suggest that selenium supplementation could aid in preventing the onset of gastric carcinogenesis in chronically infected individuals and reduce mortality in those who already have gastric ulcer [37]. Furthermore, one study indicates a correlation between selenium status and location of gastric cancer [38]. Additional research is needed to investigate why selenium levels drop before carcinogenesis and the mechanisms behind this occurrence.

3.4 Vibrio species

Selenium has demonstrated potential in combating infections caused by Vibrio species, such as *Vibrio cholerae* and *V. parahaemolyticus.* These pathogenic bacteria cause toxin-mediated diarrhea and seafood-related gastroenteritis in humans, respectively, and can lead to severe dehydration and even death in untreated patients. Innovative strategies to control and prevent such infections are necessary for enhanced public health.

Selenium has been shown to reduce *V. cholerae*'s motility, intestinal cell attachment, and cholera toxin production. The reduction in motility, an essential step in the pathogenesis of *V. cholerae,* may be due to alterations in membrane integrity that affect flagellar structure. These findings suggest that selenium supplementation can benefit the host by enhancing their immune response, while simultaneously decreasing the virulence of the bacterial pathogen [39].

Biogenic selenium nanoparticles stabilized using seaweed have exhibited significant antibacterial activity against *V. parahaemolyticus.* Scanning electron microscopy analysis revealed that the nanoparticles interact with the bacterium, attaching to the cell membrane and causing non-viability [40]. Similarly, selenium nanoparticles synthesized from marine macroalgae have demonstrated antimicrobial activity against pathogenic *V. harveyi* and *V. parahaemolyticus* [41]. This finding suggests the potential applicability of these nanoparticles in combating a broader range of Vibrio species in aquaculture.

3.5 *Clostridioides difficile*

Clostridioides difficile is a pathogenic bacterium causing toxin-mediated enteric disease in humans, mainly affecting hospital inpatients and the elderly undergoing prolonged antibiotic therapy. The rise of hypervirulent strains has resulted in *C. difficile* being listed as one of three urgent threats to human health. Although antibiotics are the drug of choice for treating *C. difficile* infections, the emergence of antibiotic resistance has led to the investigation of alternative treatments. The use of sodium selenite as an alternative therapeutic agent was shown to reduce the virulence of *C. difficile* by reducing exotoxin production without affecting the growth of beneficial bacteria commonly found in the human gastrointestinal tract. Furthermore, sodium selenite significantly increased the sensitivity of *C. difficile* to ciprofloxacin [42].

3.6 *Acinetobacter baumannii*

Acinetobacter baumannii is a multidrug-resistant pathogen that causes wound infections in humans. Due to its ability to form biofilms and colonize epithelial cells, *A. baumannii* infections can be difficult to treat. A study exploring the potential of selenium in inhibiting *A. baumannii*'s ability to form biofilms and colonize human skin keratinocytes was found to reduce bacterial adhesion and invasion of human skin keratinocytes, disrupt biofilm architecture, and downregulate genes associated with biofilm production [43].

3.7 Selenium nanoparticles for bacterial infections

Selenium nanoparticles (SeNPs) have garnered attention for their unique physicochemical properties, which include size, surface charge, and concentration, all of which influence their antimicrobial activity. The differential antimicrobial effects of SeNP on Gram-positive and Gram-negative bacteria, as well as fungi like *Candida species*, have been explored in several studies. For example, SeNPs synthesized by *Providencia vermicola* BGRW exhibited a strong inhibitory effect on the growth of several Gram-positive pathogens (such as *S. aureus*, *B. cereus*, methicillin-resistant *S. aureus*, and *Streptococcus agalactiae*) and *E. coli*, but most Gram-negative bacteria and *Candida albicans* were not inhibited [44].

The surface charge of SeNPs, which can be either positive or negative depending on the synthesis method, affects their interaction with bacterial cells. Studies have shown that negatively charged nanoparticles exhibit higher antimicrobial activity against Gram-positive bacteria due to electrostatic attraction between the negatively charged nanoparticles and the positively charged bacterial cell surface [14]. On the other hand, negatively charged SeNPs do not exhibit the same effect on Gram-negative bacteria, as the small size of penetration channels in their cell walls and the insufficient negatively charged regions on the cell wall hinder the attachment of positively charged SeNPs.

SeNPs also exhibit potential as an antimicrobial agent in combination with conventional antibiotics. By increasing the bioavailability of these agents and reducing the likelihood of antibiotic resistance, SeNPs can enhance the effectiveness of existing treatments. For instance, Menon et al. [45] demonstrated that *Klebsiella sp*. was the most susceptible to SeNP administration at a concentration of 100 μg/ml, with Serratia sp. and *S. aureus* also exhibiting significant growth reduction. SeNPs can be produced by lactic acid bacteria at ambient temperatures and pressures, providing a cost-effective and environmentally friendly alternative to chemically based methods [46].

3.7.1 Selenium nanoparticles against foodborne pathogens

The biosynthesized SeNP from *Bacillus licheniformis* has been shown to effectively control growth and biofilm formation of foodborne pathogens such as *B. cereus*, *Enterococcus faecalis*, *E. coli* O157 H7, *S. aureus*, *Salmonella Typhimurium*, and *S. enteritidis*. Although they did not completely remove established biofilms, a concentration of 75 mg/ml showed a slight effect, and SeNPs demonstrated no toxicity on Artemia larvae, making them a promising agent for preventing biofilm formation by foodborne pathogens [47].

3.7.2 Selenium nanoparticles against Pseudomonas aeruginosa

The antibacterial activity of SeNP synthesized by *Stenotrophomonas maltophilia* and *Bacillus mycoides* was assessed against clinical isolates of *Pseudomonas aeruginosa*. These

SeNPs demonstrated inhibitory effects on bacterial growth at concentrations ranging from 8 to 512 mg/ml. Conversely, the SeNP displayed no inhibitory activity against *Candida albicans* and *Candida parapsilosis* species [48]. These findings suggest that antibacterial activity of SeNP may be bacterium specific. Consequently, researchers have sought to optimize the physicochemical properties of SeNP, such as stabilization and interaction with biological molecules to broaden their spectrum of antimicrobial activity.

3.8 Selenium interactions with antimicrobials

SeNPs have demonstrated promising synergistic activity when combined with other antimicrobials. In a study that explored the potential synergistic effects, SeNPS were generated using a simple wet chemical method and combined with a set concentration of lysozyme, creating a nanohybrid system incorporating both SeNPs and lysozyme. Antibacterial tests were conducted on *S. aureus* and *E. coli,* revealing that SeNPs played a crucial role in inhibiting bacterial growth at very low protein concentrations. Furthermore, individual nanoparticles effectively suppressed bacterial growth even in the presence of high lysozyme concentrations when used in the modest amounts [49].

Huang et al. developed a synergistic nanocomposite by conjugating quercetin and acetylcholine to the surface of SeNPs, which are synthesized by chemically reducing Na2SeO3 [50]. According to their findings, the nanoparticles interacted with the bacterial cell wall, causing permanent damage to the membrane, and exhibiting remarkable synergistic antibacterial activity against MRSA at low doses. The results suggest that the synergistic effects of quercetin and acetylcholine increase the antibacterial activity of SeNPs [50].

Cihalova et al. reported that SeNPs possess potent inhibitory action when combined with conventional antibiotics. Using an impedance method, they observed a greater disruption of biofilms after applying antibiotic complexes containing SeNPs compared to those treated with antibiotics alone. In comparison with bacteria without antibacterial compounds, the nanoparticles inhibited the formation of MRSA biofilms by up to 94% ± 4%, while drugs without SeNPs only suppress MRSA by up to 16% ± 2% [51]. This evidence highlights the potential for SeNPs to enhance the efficacy of antimicrobial treatments through synergistic interactions with other antimicrobials.

4. Antiviral activity of selenium and selenium nanoparticles

Beyond its involvement in bacterial infections, selenium has also been implicated in viral infections. Studies have indicated that selenium deficiency can exacerbate the pathogenicity of certain viruses, while adequate selenium levels contribute to improved immune responses and viral clearance. Selenium is vital in defending the host system against viral infections in various infectious diseases. Nutritional deficiencies in selenium can affect both the pathogenicity of a virus and the immune system's response [52]. Selenium compounds, such as selenite, can inhibit viral invasion of healthy cells and reduce their infectiousness [53]. Moreover, selenium and vitamin E supplements have shown to increase resistance to respiratory viral infections [3].

4.1 SARS-CoV-2

The COVID-19 pandemic, caused by severe acute respiratory syndrome corona virus 2 (SARS-CoV-2), emerged in 2019 and has globally affected about 530 million people, causing 6.3 million deaths [54]. Selenium may protect the host due to its critical role as a cofactor for enzymes that work with vitamin E to reduce the generation of ROS, which can cause oxidative damage in both pathogen and host cells [55]. The main SARS-CoV-2 protease interacts with glutathione peroxidase1 (GPX1), a crucial selenium-dependent enzyme responsible for viral replication [56]. Notably, the GPX1 mimic synthetic selenium compound ebselen is a potent inhibitor of SARS-CoV-2 virus main protease enzyme [57].

Sodium selenite can oxidize the thiol groups on the surface of the coronavirus protein disulfide isomerase, preventing it from penetrating healthy cell membranes. Wang et al. demonstrated the potential of SeNPs for COVID-19 diagnosis using a lateral flow immunoassay kit based on SeNPs-modified SARS-CoV-2 nucleoprotein, which detected anti-SARS-CoV-2 IgM and IgG in human serum within 10 minutes with the naked eye [58].

4.2 Human immunodeficiency virus

Human immunodeficiency virus (HIV) is an RNA virus in the Lentivirus genus that causes acquired immunodeficiency syndrome (AIDS), leading to a compromised immune system by infecting immune cells. HIV currently affects over 37 million individuals and causes 1.5 million annual deaths [59]. Selenium has been shown to suppress HIV *in vitro* due to its antioxidant properties as a component of GPx and other selenoproteins. Many studies have reported low serum selenium levels in HIV-positive individuals, and serum selenium levels decrease as the disease progresses. Several cohort studies have established a link between selenium deficiency and the development of AIDS. Although some randomized controlled trials have shown that selenium supplementation can improve CD4+ cell counts and reduce hospitalizations and diarrheal morbidity, additional follow-up studies are needed to confirm this finding [60].

4.3 Influenza virus

Influenza virus affects the respiratory tract, and acute pneumonia is diagnosed in 30–40% of hospitalized individuals with laboratory-confirmed influenza. Influenza A is the most common viral cause of acute respiratory distress syndrome (ARDS) in adults [61]. Selenium therapy has been shown to modify the response to the influenza vaccination in older adults, which was associated with elevated IFN-γ levels following vaccination [62, 63]. Li et al. developed oseltamivir adorned SeNPs to treat the H1N1 virus. These compounds significantly hindered the H1N1 influenza virus's ability to bind to host cells by preventing the activities of hemagglutinin and neuraminidase [13, 64]. SeNPs have demonstrated potential in combating the H1N1 influenza virus by blocking the ROS-mediated AKT and p53 signaling pathways, thereby preventing apoptosis, DNA fragmentation, chromatin condensation, and ultimately, cell death [28, 65]. Moreover, SeNPs can prevent cellular and lung tissue damage caused by the H1N1 virus [66]. Studies in broiler chickens revealed that while hexanic extracts of fig and olive fruit, along with nano-selenium, induced some immunity against the H9N2 avian influenza virus, they were unable to prevent anamnestic reactions or infections [67]. Research by Shojadoost et al. indicated that selenium supplementation enhances the immunity provided by vaccines, as shown by increased antibody levels (IgM and IgY) and reduced virus shedding in chickens treated with organic and inorganic selenium [68]. In mice, a ruthenium-selenium metal complex exhibited antiviral mechanisms by inhibiting viral assembly and replication, controlling virus-mediated apoptosis, and reducing lung tissue inflammation [69].

4.4 Acute respiratory distress syndrome

Acute respiratory distress syndrome (ARDS), a common cause of respiratory failure in critically ill patients, is characterized by noncardiogenic pulmonary edema, hypoxemia, and mechanical ventilation requirements [70]. A case study investigated the impact of sodium selenite on ARDS and found that patients treated with it for 10 days experienced reduced airway resistance, improved lung compliance, increased fraction of inspired oxygen (FiO2), higher arterial oxygen pressure (PaO2), shorter hospital stays, and lower mortality rates. Selenium supplementation was found to restore lung antioxidant capacity, regulate inflammatory responses *via* interleukin (IL)-1 and IL-6 levels, and significantly enhance respiratory mechanics [71, 72].

4.5 Hepatitis virus

Viral hepatitis, which causes over 1.3 million deaths annually worldwide, is a major global health concern [73]. Although current anti-HIV drugs help control the epidemic, side effects and drug resistance call for safer and more effective treatment options. Sodium selenite has been found to inhibit Hepatitis B virus (HBV) protein expression, transcription, and genome replication in hepatoma cell cultures in a dose-dependent manner [13, 74]. By administering SeNPs and the hepatitis B antigen vaccination, Mahdavi et al. devised a method that could increase IFN-g levels, stimulate a Th1 response, and thus improve vaccine efficacy by activating the immune system toward a Th1 state [75].

4.6 Enterovirus

Enterovirus 71 (EV71) is the primary pathogen responsible for severe cases of hand, foot, and mouth disease (HFMD), for which there is currently no effective treatment [76]. Oseltamivir, a potent antiviral drug, was loaded onto SeNPs to enhance its antiviral activity against EV71. The functionalized SeNPs improved oseltamivir's efficacy by inhibiting EV71 growth, preventing cell death, and reducing caspase-3 activity and ROS generation [76]. Additionally, SeNPs were used to load small interfering RNA (siRNA) targeting the EV71 Vp1 gene, with polyethylenimine (PEI) decorating the surface (Se@PEI@siRNA). In nerve cell line, Se@PEI@siRNA demonstrated high interference efficiency and protected cells from infection [77].

5. Antifungal activity of selenium and selenium nanoparticles

Selenium has emerged as a promising agent in mitigating the harmful effects of mycotoxins such as aflatoxin B1 (AFB1) and ochratoxin A (OTA), which pose significant health risks and economic losses due to their prevalence in foods. Further, SeNPs have recently gained interest for their superior antifungal properties and ability to inhibit the growth of multidrug-resistant fungus, offering potential strategies against mycotoxin-induced health issues.

AFB1 is a potent mycotoxin produced by certain strains of *Aspergillus* fungi (such as *Aspergillus flavus* and *Aspergillus parasiticus*), and is a prevalent contaminant in food, contributing to health issues in humans. Chronic exposure to AFB1 has been associated with immune toxicity, carcinogenicity, genotoxicity, hepatotoxicity, and reproductive disorders. AFB1 undergoes bioactivation in the liver to a highly reactive form exo-AFB1–8,9-epoxide (AFBO) that can cause DNA damage. Selenium-fortified yogurt has been shown to mitigate the harmful effects of aflatoxins in mice, such as weight loss and reduced food intake, by enhancing aflatoxin detoxification pathways and preventing AFB1-DNA adduct formation [78]. AFB1 can also trigger oxidative stress by generating ROS, potentially necessitating cytochromeP450 (CYP450) activation. Dietary selenium was shown to mitigate AFB1-induced liver damage in chickens by inhibiting CYP450 activation of AFB1 and enhancing antioxidant responses through selenoprotein gene upregulation [79]. AFB1 has also been reported to impair immune function, increasing susceptibility to infectious diseases. However, selenium supplementation, especially in the form of organic selenium, selenomethionine (SeMet), has demonstrated promising results in ameliorating AFB1-induced immune toxicity. The protective effects of SeMet were largely attributed to its ability to boost the expression of GPx1 and selenoprotein S, key element in antioxidant defense [80]. Selenium has also been the subject of extensive research due to its potential role in activating testosterone synthesis. Research has demonstrated the protective effects of selenium against AFB1-induced testicular toxicity. Specifically, selenium was found to improve testes index, sperm functional parameters (including concentration, malformation, and motility), and serum testosterone levels in AFB1-exposed mice. These findings suggest that selenium can effectively mitigate the oxidative stress and impaired testosterone synthesis induced by AFB1 exposure [81].

Kashin-Beck disease (KBD) characterized by severe osteoarthritis has been associated with low environmental selenium and the involvement of mycotoxins. A study conducted by Hong et al. has shown that selenium influences the growth of *Fusarium* strains and decreases chondrocyte injury indicators when chondrocytes are exposed to extracts from these fungal cultures. These findings suggest a link among environmental selenium levels, fungal metabolite production, and chondrocyte damage, which warrants further exploration [82].

Ochratoxin A, a mycotoxin produced by *Penicillium* and *Aspergillus* molds, poses significant health risks due to its widespread presence in crops and its ability to cause kidney and liver lesions, immune dysfunction, and genotoxicity in humans and animals. The exact mechanism of OTA's toxicity, which has been linked to oxidative stress and cytotoxicity, remains under investigation [83, 84]. However, recent research suggests that selenium may counteract OTA's cytotoxicity and oxidative stress damage. Various studies have shown that selenium can enhance cell survival after OTA exposure, activate the antioxidant response, and reduce oxidative stress and apoptosis in OTA-induced kidney injury [85]. Both SeMet and sodium selenite have demonstrated protective effects, possible through upregulation of antioxidant enzyme expression and the downregulation of apoptosis-related factors [86]. In combination with zinc, selenium was found to alleviate ochratoxin A-induced fibrosis in human kidney cells by blocking ROS dependent autophagy offering a new perspective on nutritional interventions against mycotoxin-induced health issues [87].

More recently, the role of biosynthesized selenium nanoparticles has gained attention due to their enhanced antifungal properties. Studies have shown that SeNP, biosynthesized using plant extracts or *Aspergillus oryzae* fermented lupin extract, can effectively inhibit the growth of multidrug-resistant bacteria and pathogenic fungi [88]. These nanoparticles have also demonstrated an effect on the expression of CYP51A and HSP90 antifungal resistance genes in *Ammophilus fumigatus* and *A. flavus* [89]. In *Candida albicans* isolates, biogenic SeNP was found to reduce the expression of ERG11 and CDR1 genes that are associated with azole resistance [90]. Furthermore, when compared to gold and silver nanoparticles, SeNPs exhibited superior antifungal properties against amphotericin B-resistant *Candida glabrata* clinical isolates [91]. Furthermore, the capacity of biogenic SeNPs to disrupt biofilms, particularly those formed by *C. albicans,* a primary causative agent of hospital-related infections stemming from biofilms on medical devices, has also been effectively demonstrated [92] without being cytotoxic to human embryonic kidney cells, thereby highlighting their potential as safe and efficacious agent in combating such infections.

To summarize, selenium, whether in its organic form or as biosynthesized nanoparticles, displays significant antifungal properties. By mitigating mycotoxin-induced toxicity and inhibiting the growth of various fungal species, selenium serves as a potential candidate for the development of novel antifungal strategies.

6. Limitations and toxicity of selenium

While selenium is an essential micronutrient with numerous health benefits, its toxicity and potential adverse effects must be considered. The toxicity of selenium depends on its chemical form, with organic selenium compounds generally being less harmful than their inorganic counterparts. However, the lethal dose (LD50) values can vary significantly based on the duration of exposure, the model employed, and the blood levels reached [93].

Recent studies have shown that intravenous administration of sodium selenite at a dose of 500 μg/day is non-toxic [94], and even relatively high dosages (up to 2000 μg/day) were well tolerated in individuals with peritonitis [95]. Nevertheless, excessively high selenium blood levels (>1 mg/L) can lead to selenosis, a condition characterized by gastrointestinal disturbances, hair loss, white blotchy nails, garlic breath odor, fatigue, irritability, and mild nerve damage [96]. The sodium selenite LD50 dose for rats is 4100 μg/kg body weight, which is 100 times higher than the dose typically used in humans [53]. In human serum, selenium concentrations range from 400 to 3000 μg/L, with levels above 1400 μg/L being non-toxic [93]. It is generally believed, though not definitively proven, that hazardous levels of selenite begin at 600 μg/day [53].

Given the potential toxicity of selenium at high doses, it is crucial to control the therapeutic dose. Plant-based nanoparticles may help mitigate the harmful effects of selenium, as they have been found to be less toxic than inorganic selenium [97]. Various physical and chemical methods have been employed to produce SeNPs, involving the use of different chemical compounds and physical processes. However, the high cost of these technologies and the potential contamination of nanoparticles with harmful chemical residues limit SeNPs therapeutic application in the pharmaceutical and medical industries [97]. As research into selenium's antimicrobial properties continues, it is essential to maintain a balance between its therapeutic benefits and potential adverse effects.

6.1 Microbial resistance to selenium

Although selenium has demonstrated antimicrobial properties, the potential for microbial resistance to selenium remains an area that requires further investigation. Researchers have predominantly focused on the reduction of selenium to less toxic or harmless SeNPs and methylated selenium, but not all bacteria can reduce toxic oxyanions, and the resulting selenium species may not be methylated [98]. Moreover, there is currently limited information on the mechanisms of selenium resistance in bacteria, such as efflux and sorption of selenium oxyanions [98].

Interestingly, when use at the nanoscale, SeNPs have been shown to inhibit the dissemination of environmental antibiotic resistance genes, providing effective antibacterial properties without complicating the scale-up harvesting process [99]. As research into selenium's antimicrobial potential continues, it is crucial to expand our understanding of microbial resistance mechanisms to ensure the effective and sustainable use of selenium-based treatments.

6.2 Variability in selenium availability in different population

Variability in selenium intake across the globe is influenced by several factors, including selenium concentration in soil, as well as factors affecting its availability in the food chain, such as the type of selenium, soil pH, organic matter content, and the presence of ions [100]. Most of Europe has lower selenium content in the soil compared to the United States, with Eastern Europe having a lower selenium intake than Western Europe. It is estimated that 15% of the global population experiences selenium deficiency, and selenium intake varies significantly between countries. Dietary selenium intake is approximately 40 µg per day in Europe, while in the USA, daily selenium intake ranges from 93 µg/day in women to 134 µg/day in men [101, 102].

Considering gender differences, the recommended daily selenium allowance in the United Kingdom is 75 µg/day for men and 60 µg/day for women [103]. This variability in selenium intake across different regions can lead to deficiency-related diseases in areas where intake is insufficient. Consequently, populations in these areas become more vulnerable to infectious disease due to the inadequate selenium consumption.

7. Conclusions

Selenium has been demonstrated to possess antimicrobial properties against various public health pathogens. In addition, its potential to modulate immune responses, generate ROS, and disrupt microbial processes highlights its importance in the fight against infectious diseases. Despite these promising findings, challenges remain, such as bioavailability, toxicity, and development of microbial resistance. Overcoming these obstacles necessitates further research, collaboration, and well-designed clinical trials. As we deepen our understanding and develop innovative solutions, selenium may emerge as a vital addition to our arsenal of antimicrobial agents, playing a crucial role in safeguarding public health, especially in light of rising antimicrobial resistance.

DOI: http://dx.doi.org/10.5772/intechopen.111879

Author details

Poonam Gopika Vinayamohan, Divya Joseph, Leya Susan Viju
and Kumar Venkitanarayanan*
Department of Animal Science, University of Connecticut, Storrs, USA

*Address all correspondence to: kumar.venkitanarayanan@uconn.edu

References

[1] Ge K, Xue A, Bai J, Wang S. Keshan disease-an endemic cardiomyopathy in China. Virchows Archiv A. 1983;**401**(1):1-15

[2] Goldhaber SB. Trace element risk assessment: Essentiality vs. toxicity. Regulatory Toxicology and Pharmacology. 2003;**38**(2):232-242

[3] Shakoor H, Feehan J, Al Dhaheri AS, Ali HI, Platat C, Ismail LC, et al. Immune-boosting role of vitamins D, C, E, zinc, selenium and omega-3 fatty acids: Could they help against COVID-19? Maturitas. 2021;**143**:1-9

[4] Baker RD, Baker SS, LaRosa K, Whitney C, Newburger PE. Selenium regulation of glutathione peroxidase in human hepatoma cell line Hep3B. Archives of Biochemistry and Biophysics. 1993;**304**(1):53-57. DOI: 10.1006/abbi.1993.1320

[5] Shojadoost B, Kulkarni RR, Yitbarek A, Laursen A, Taha-Abdelaziz K, Negash Alkie T, et al. Dietary selenium supplementation enhances antiviral immunity in chickens challenged with low pathogenic avian influenza virus subtype H9N2. Veterinary Immunology and Immunopathology. 2019;**207**:62-68

[6] Fairweather-Tait SJ, Bao Y, Broadley MR, Collings R, Ford D, Hesketh JE, Hurst R. Selenium in Human Health and Disease. Antioxidants & Redox Signaling. 2011;**14**(7):1337-1383. DOI: 10.1089/ars.2010.3275

[7] Huang Z, Rose AH, Hoffmann PR. The Role of Selenium in Inflammation and Immunity: From Molecular Mechanisms to Therapeutic Opportunities. Antioxidants & Redox Signaling. 2012;**16**(7):705-743. DOI: 10.1089/ars.2011.4145

[8] Gandhi UH, Kaushal N, Ravindra KC, Hegde S, Nelson SM, Narayan V, et al. Selenoprotein-dependent up-regulation of hematopoietic prostaglandin D2 synthase in macrophages is mediated through the activation of peroxisome proliferator-activated receptor (PPAR) gamma. The Journal of Biological Chemistry. 2011;**286**(31):27471-27482. DOI: 10.1074/jbc.M111.260547

[9] Steinbrenner H, Al-Quraishy S, Dkhil MA, Wunderlich F, Sies H. Dietary selenium in adjuvant therapy of viral and bacterial infections. Advances in Nutrition. 2015;**6**(1):73-82

[10] Cui D, Ma J, Liang T, Sun L, Meng L, Liang T, et al. Selenium nanoparticles fabricated in laminarin polysaccharides solutions exert their cytotoxicities in HepG2 cells by inhibiting autophagy and promoting apoptosis. International Journal of Biological Macromolecules. 2019;**137**:829-835. DOI: 10.1016/j.ijbiomac.2019.07.031

[11] Neelapu SS, Tummala S, Kebriaei P, Wierda W, Gutierrez C, Locke FL, et al. Chimeric antigen receptor T-cell therapy—Assessment and management of toxicities. Nature Reviews. Clinical Oncology. 2018;**15**(1):47-62. DOI: 10.1038/nrclinonc.2017.148

[12] Xia IF, Cheung JS, Wu M, Wong KS, Kong HK, Zheng XT, et al. Dietary chitosan-selenium nanoparticle (CTS-SeNP) enhance immunity and disease resistance in zebrafish. Fish & Shellfish Immunology. 2019;**87**:449-459

[13] Lin W, Zhang J, Xu JF, Pi J. The advancing of selenium nanoparticles

against infectious diseases. Frontiers in Pharmacology. 2021;**12**:682284 [Internet]. [cited 2023 May 11] Available from: https://www.frontiersin.org/articles/10.3389/fphar.2021.682284

[14] Sumner SE, Markley RL, Kirimanjeswara GS. Role of Selenoproteins in bacterial pathogenesis. Biological Trace Element Research. 2019;**192**(1):69-82

[15] Meziane W, Habi S, Boulatika Y, Marchandin H, Aymeric JL. Macrophage bactericidal activities against Staphylococcus aureus are enhanced in vivo by selenium supplementation in a dose-dependent manner. PLoS One. 2015;**9**:10-18 [Internet] [cited 2023 May 11]; Available from: http://prodinra.inra.fr/ft/6A7E3A29-B0CF-49A1-BCDA-DC24DF57E901

[16] Bi CL, Wang H, Wang YJ, Sun J, Dong JS, Meng X, et al. Selenium inhibits Staphylococcus aureus-induced inflammation by suppressing the activation of the NF-κB and MAPK signalling pathways in RAW264.7 macrophages. European Journal of Pharmacology. 2016;**780**:159-165

[17] Wei Z, Yao M, Li Y, He X, Yang Z. Dietary selenium deficiency exacerbates lipopolysaccharide-induced inflammatory response in mouse mastitis models. Inflammation. 2014;**37**(6):1925-1931

[18] Malbe M, Klaassen M, Fang W, Myllys V, Vikerpuur M, Nyholm K, et al. Comparisonsofseleniteandseleniumyeast feed supplements on Se-incorporation, mastitis and leucocyte function in Se-deficient dairy cows. Journal of Veterinary Medicine Series A. 1995;**42**(1-10):111-121

[19] Gao X, Zhang Z, Li Y, Hu X, Shen P, Fu Y, et al. Selenium deficiency deteriorate the inflammation of S. aureus infection via regulating NF-κB and PPAR-γ in mammary gland of mice. Biological Trace Element Research. 2016;**172**(1):140-147

[20] Liu Y, Gao Y, Liu X, Liu Q, Zhang Y, Wang Q, et al. Transposon insertion sequencing reveals T4SS as the major genetic trait for conjugation transfer of multi-drug resistance pEIB202 from Edwardsiella. BMC Microbiology. 2017;**17**(1):112

[21] Full Article: Selenium Nanoparticles Inhibit *Staphylococcus aureus* Growth [Internet]. [cited 2023 May 11]. Available from: https://www.tandfonline.com/doi/full/10.2147/IJN.S21729

[22] Sonkusre P, Singh CS. Biogenic selenium nanoparticles inhibit Staphylococcus aureus adherence on different surfaces. Colloids and Surfaces. B, Biointerfaces. 2015;**136**:1051-1057

[23] Golmohammadi R, Najar-Peerayeh S, Tohidi Moghadam T, Hosseini SMJ. Synergistic antibacterial activity and wound healing properties of selenium-chitosan-Mupirocin Nanohybrid system: An in vivo study on rat diabetic Staphylococcus aureus wound infection model. Scientific Reports. 2020;**10**(1):2854

[24] Smith AD, Botero S, Shea-Donohue T, Urban JF. The pathogenicity of an enteric Citrobacter rodentium infection is enhanced by deficiencies in the antioxidants selenium and vitamin E. Infection and Immunity. 2011;**79**(4):1471-1478

[25] Yang J, Huang K, Qin S, Wu X, Zhao Z, Chen F. Antibacterial action of selenium-enriched probiotics against pathogenic Escherichia coli. Digestive Diseases and Sciences. 2009;**54**(2):246-254

[26] Kim HW, Ha US, Woo JC, Kim SJ, Yoon BI, Cho YH, et al. Preventive effect of selenium on chronic bacterial prostatitis. Journal of Infection and Chemotherapy. 2012;**18**(1):30-34

[27] Narayanan A, Nair MS, Muyyarikkandy MS, Amalaradjou MA. Inhibition and inactivation of Uropathogenic Escherichia coli biofilms on urinary catheters by sodium selenite. International Journal of Molecular Sciences. 2018;**19**(6):1703

[28] Nair MS, Upadhyay A, Fancher S, Upadhyaya I, Dey S, Kollanoor-Johny A, et al. Inhibition and inactivation of Escherichia coli O157:H7 biofilms by selenium. Journal of Food Protection. 2018;**81**(6):926-933

[29] Surendran-Nair M, Kollanoor-Johny A, Ananda-Baskaran S, Norris C, Lee JY, Venkitanarayanan K. Selenium reduces enterohemorrhagic Escherichia coli O157:H7 verotoxin production and globotriaosylceramide receptor expression on host cells. Future Microbiology. 2016;**11**:745-756

[30] Vasi S. Influence of sodium selenite on the growth of selected bacteria species and their sensitivity to antibiotics.

[31] Kamboj AK, Cotter TG, Oxentenko AS. Helicobacter pylori: The past, present, and future in management. Mayo Clinic Proceedings. 2017;**92**(4):599-604

[32] Lahner E, Persechino S, Annibale B. Micronutrients (other than iron) and helicobacter pylori infection: A systematic review. Helicobacter. 2012;**17**(1):1-15

[33] Plasma and Gastric Tissue Selenium Levels in Patients with H... : Journal of Clinical Gastroenterology [Internet]. [cited 2023 May 11]. Available from: https://journals.lww.com/jcge/fulltext/2001/05000/plasma_and_gastric_tissue_selenium_levels_in.9.aspx

[34] Hu A, Li L, Hu C, Zhang D, Wang C, Jiang Y, et al. Serum concentrations of 15 elements among helicobacter pylori-infected residents from Lujiang County with high gastric cancer risk in eastern China. Biological Trace Element Research. 2018;**186**(1):21-30

[35] Camargo MC, Burk RF, Bravo LE, Piazuelo MB, Hill KE, Fontham ET, et al. Plasma selenium measurements in subjects from areas with contrasting gastric cancer risks in Colombia. Archives of Medical Research. 2008;**39**(4):443-451

[36] Sjunnesson H, Sturegård E, Willen R, Wadström T. High intake of selenium, beta-carotene, and vitamins a, C, and E reduces growth of helicobacter pylori in the Guinea pig. Comparative Medicine. 2001;**51**(5):418-423

[37] Cai X, Wang C, Yu W, Fan W, Wang S, Shen N, et al. Selenium exposure and cancer risk: An updated meta-analysis and meta-regression. Scientific Reports. 2016;**6**:19213

[38] Ji JH, Shin DG, Kwon Y, Cho DH, Lee KB, Park SS, et al. Clinical correlation between gastric cancer type and serum selenium and zinc levels. Journal of Gastric Cancer. 2012;**12**(4):217-222

[39] Bhattaram V, Upadhyay A, Yin HB, Mooyottu S, Venkitanarayanan K. Effect of dietary minerals on virulence attributes of vibrio cholerae. Frontiers in Microbiology. 2017;**8**:911 [Internet]. [cited 2023 May 11] Available from: https://www.frontiersin.org/articles/10.3389/fmicb.2017.00911

[40] (PDF) Biogenic zinc Oxide, Copper Oxide and Selenium Nanoparticles:

Preparation, Characterization and Their Anti-Bacterial Activity against *Vibrio parahaemolyticus* [Internet]. [cited 2023 May 11]. Available from: https://www.researchgate.net/publication/345806486_Biogenic_zinc_oxide_copper_oxide_and_selenium_nanoparticles_preparation_characterization_and_their_anti-bacterial_activity_against_Vibrio_parahaemolyticus

[41] Rajashree R, S G, M G. Marine biomolecule mediated synthesis of selenium nanoparticles and their antimicrobial efficiency against fish and crustacean pathogens. Research Square; 2022. DOI: 10.21203/rs.3.rs-1594624/v1

[42] In Vitro Efficacy of Sodium Selenite in Reducing Toxin Production, Spore Outgrowth And Antibiotic Resistance in Hypervirulent *Clostridium difficile* | Microbiology Society [Internet]. [cited 2023 May 11]. Available from: https://www.microbiologyresearch.org/content/journal/jmm/10.1099/jmm.0.001008

[43] Surendran-Nair M, Lau P, Liu Y, Venkitanarayanan K. Efficacy of selenium in controlling Acinetobacter baumannii associated wound infections. Wound Medicine. 2019;**26**(1):100165

[44] El-Deeb B, Al-Talhi A, Mostafa N, Abou-assy R. Biological synthesis and structural characterization of selenium nanoparticles and assessment of their antimicrobial properties. American Scientific Research Journal for Engineering, Technology, and Sciences. 2018;**45**(1):135-170

[45] Investigating the Antimicrobial Activities of the Biosynthesized Selenium Nanoparticles and Its Statistical Analysis | SpringerLink [Internet]. [cited 2023 May 11]. Available from: https://link.springer.com/article/10.1007/s12668-019-00710-3

[46] Martínez FG, Moreno-Martin G, Pescuma M, Madrid-Albarrán Y, Mozzi F. Biotransformation of selenium by lactic acid bacteria: Formation of Seleno-nanoparticles and Seleno-amino acids. Frontiers in Bioengineering and Biotechnology. 2020;**8**:506 [Internet]. [cited 2023 May 11] Available from: https://www.frontiersin.org/articles/10.3389/fbioe.2020.00506

[47] Khiralla GM, El-Deeb BA. Antimicrobial and antibiofilm effects of selenium nanoparticles on some foodborne pathogens. LWT - Food Sci Technol. 2015;**63**(2):1001-1007

[48] Cremonini E, Zonaro E, Donini M, Lampis S, Boaretti M, Dusi S, et al. Biogenic selenium nanoparticles: Characterization, antimicrobial activity and effects on human dendritic cells and fibroblasts. Microbial Biotechnology. 2016;**9**(6):758-771

[49] Vahdati M, Tohidi MT. Synthesis and characterization of selenium nanoparticles-lysozyme Nanohybrid system with synergistic antibacterial properties. Scientific Reports. 2020;**10**(1):510

[50] Huang X, Chen X, Chen Q, Yu Q, Sun D, Liu J. Investigation of functional selenium nanoparticles as potent antimicrobial agents against superbugs. Acta Biomaterialia. 2016;**30**:397-407

[51] Cihalova K, Chudobova D, Michalek P, Moulick A, Guran R, Kopel P, et al. Staphylococcus aureus and MRSA growth and biofilm formation after treatment with antibiotics and SeNPs. International Journal of Molecular Sciences. 2015;**16**(10):24656-24672

[52] Moghaddam A, Heller RA, Sun Q, Seelig J, Cherkezov A, Seibert L, et al. Selenium deficiency is associated with mortality risk from COVID-19. Nutrients. 2020;**12**(7):2098

[53] Kieliszek M, Lipinski B. Selenium supplementation in the prevention of coronavirus infections (COVID-19). Medical Hypotheses. 2020;**143**:109878

[54] Nagpal D, Nagpal S, Kaushik D, Kathuria H. Current clinical status of new COVID-19 vaccines and immunotherapy. Environmental Science and Pollution Research International. 2022;**29**(47):70772-70807

[55] Harthill M. Review: Micronutrient selenium deficiency influences evolution of some viral infectious diseases. Biological Trace Element Research. 2011;**143**(3):1325-1336

[56] Seale LA, Torres DJ, Berry MJ, Pitts MW. A role for selenium-dependent GPX1 in SARS-CoV-2 virulence. The American Journal of Clinical Nutrition. 2020;**112**(2):447-448

[57] Sies H, Parnham MJ. Potential therapeutic use of ebselen for COVID-19 and other respiratory viral infections. Free Radical Biology & Medicine. 2020;**156**:107-112

[58] A Point-of-Care Selenium Nanoparticle-based Test for the Combined Detection of Anti-SARS-CoV-2 IgM and IgG in Human Serum and Blood - Lab on a Chip (RSC Publishing) [Internet]. [cited 2023 May 11]. Available from: https://pubs.rsc.org/en/content/articlelanding/2020/lc/d0lc00828a

[59] Guillin OM, Vindry C, Ohlmann T, Chavatte L. Interplay between selenium, Selenoproteins and HIV-1 replication in human CD4 T-lymphocytes. International Journal of Molecular Sciences. 2022;**23**(3):1394

[60] Stone CA, Kawai K, Kupka R, Fawzi WW. The role of selenium in HIV infection Cosby a Stone, Kosuke Kawai, Roland Kupka, Wafaie W Fawzi Harvard School of Public Health. Nutrition Reviews. 2010;**68**(11):671-681

[61] Kalil AC, Thomas PG. Influenza virus-related critical illness: Pathophysiology and epidemiology. Critical Care London England. 2019;**23**(1):258

[62] Alexander J, Tinkov A, Strand TA, Alehagen U, Skalny A, Aaseth J. Early nutritional interventions with zinc, selenium and vitamin D for raising anti-viral resistance against progressive COVID-19. Nutrients. 2020;**12**(8):2358

[63] Ivory K, Prieto E, Spinks C, Armah CN, Goldson AJ, Dainty JR, et al. Selenium supplementation has beneficial and detrimental effects on immunity to influenza vaccine in older adults. Clinical nutrition (Edinburgh, Scotland). 2017;**36**(2):407-415

[64] Li Y, Lin Z, Guo M, Xia Y, Zhao M, Wang C, et al. Inhibitory activity of selenium nanoparticles functionalized with oseltamivir on H1N1 influenza virus. International Journal of Nanomedicine. 2017;**12**:5733-5743

[65] Li Y, Lin Z, Guo M, Zhao M, Xia Y, Wang C, et al. Inhibition of H1N1 influenza virus-induced apoptosis by functionalized selenium nanoparticles with amantadine through ROS-mediated AKT signaling pathways. International Journal of Nanomedicine. 2018;**13**:2005-2016

[66] Lin Z, Li Y, Gong G, Xia Y, Wang C, Chen Y, et al. Restriction of H1N1 influenza virus infection by selenium nanoparticles loaded with ribavirin via resisting caspase-3 apoptotic pathway. International Journal of Nanomedicine. 2018;**13**:5787-5797

[67] Asl Najjari AH, Rajabi Z, Vasfi Marandi M, Dehghan G. The effect of the hexanic extracts of fig (Ficus

carica) and olive (Olea europaea) fruit and nanoparticles of selenium on the immunogenicity of the inactivated avian influenza virus subtype H9N2. Veterinary Research Forum International Journal. 2015;**6**(3):227-231

[68] Shojadoost B, Taha-Abdelaziz K, Alkie TN, Bekele-Yitbarek A, Barjesteh N, Laursen A, et al. Supplemental dietary selenium enhances immune responses conferred by a vaccine against low pathogenicity avian influenza virus. Veterinary Immunology and Immunopathology. 2020;**227**:110089

[69] Li Y, Chen D, Su J, Chen M, Chen T, Jia W, et al. Selenium-ruthenium complex blocks H1N1 influenza virus-induced cell damage by activating GPx1/TrxR1. Theranostics. 2023;**13**(6):1843-1859

[70] Matthay MA, Zemans RL, Zimmerman GA, Arabi YM, Beitler JR, Mercat A, et al. Acute respiratory distress syndrome. Nature Reviews Disease Primers. 2019;**5**(1):18

[71] Mahmoodpoor A, Hamishehkar H, Shadvar K, Ostadi Z, Sanaie S, Saghaleini SH, et al. The effect of intravenous selenium on oxidative stress in critically ill patients with acute respiratory distress syndrome. Immunological Investigations. 2019;**48**(2):147-159

[72] Oliveira CR, Viana ET, Gonçalves TF, Mateus-Silva JR, Vieira RP. Therapeutic use of intravenous selenium in respiratory and immunological diseases: A narrative review. Advances in Respiratory Medicine. 2022;**90**(2):134-142

[73] Rasche A, Sander AL, Corman VM, Drexler JF. Evolutionary biology of human hepatitis viruses. Journal of Hepatology. 2019;**70**(3):501-520

[74] Cheng Z, Zhi X, Sun G, Guo W, Huang Y, Sun W, et al. Sodium selenite suppresses hepatitis B virus transcription and replication in human hepatoma cell lines. Journal of Medical Virology. 2016;**88**(4):653-663

[75] Mahdavi M, Mavandadnejad F, Yazdi MH, Faghfuri E, Hashemi H, Homayouni-Oreh S, et al. Oral administration of synthetic selenium nanoparticles induced robust Th1 cytokine pattern after HBs antigen vaccination in mouse model. Journal of Infection and Public Health. 2017;**10**(1):102-109

[76] Zhong J, Xia Y, Hua L, Liu X, Xiao M, Xu T, et al. Functionalized selenium nanoparticles enhance the anti-EV71 activity of oseltamivir in human astrocytoma cell model. Artificial Cells, Nanomedicine, and Biotechnology. 2019;**47**(1):3485-3491

[77] Lin Z, Li Y, Xu T, Guo M, Wang C, Zhao M, et al. Inhibition of enterovirus 71 by selenium nanoparticles loaded with siRNA through Bax Signaling pathways. ACS Omega. 2020;**5**(21):12495-12500

[78] Alsuhaibani AMA. Functional role of selenium-fortified yogurt against aflatoxin-contaminated nuts in rats. Agriculture and Food Security. 2018;**7**(1):21

[79] Sun LH, Zhang NY, Zhu MK, Zhao L, Zhou JC, Qi DS. Prevention of aflatoxin B1 Hepatoxicity by dietary selenium is associated with inhibition of cytochrome P450 isozymes and up-regulation of 6 Selenoprotein genes in Chick liver. The Journal of Nutrition. 2015;**146**(4):655-661

[80] Hao S, Hu J, Song S, Huang D, Xu H, Qian G, et al. Selenium alleviates aflatoxin B_1-induced immune toxicity through improving glutathione peroxidase 1 and Selenoprotein

S expression in primary porcine Splenocytes. Journal of Agricultural and Food Chemistry. 2016;**64**(6):1385-1393

[81] Cao Z, Shao B, Xu F, Liu Y, Li Y, Zhu Y. Protective effect of selenium on aflatoxin B1-induced testicular toxicity in mice. Biological Trace Element Research. 2017;**180**(2):233-238

[82] Yin H, Zhang Y, Zhang F, Hu JT, Zhao YM, Cheng BL. Effects of selenium on Fusarium growth and associated fermentation products and the relationship with chondrocyte viability. Biomedicine Environment Science BES. 2017;**30**(2):134-138

[83] Gautier JC, Holzhaeuser D, Markovic J, Gremaud E, Schilter B, Turesky RJ. Oxidative damage and stress response from ochratoxin a exposure in rats. Free Radical Biology & Medicine. 2001;**30**(10):1089-1098

[84] Mally A, Zepnik H, Wanek P, Eder E, Dingley K, Ihmels H, et al. Ochratoxin a: Lack of formation of covalent DNA adducts. Chemical Research in Toxicology. 2004;**17**(2):234-242

[85] Zhang Z, Sun Y, Xie H, Wang J, Zhang X, Shi Z, et al. Protective effect of selenomethionine on kidney injury induced by ochratoxin a in rabbits. Environmental Science and Pollution Research. 2023;**30**(11):29874-29887

[86] Ren Z, He H, Fan Y, Chen C, Zuo Z, Deng J. Research Progress on the toxic antagonism of selenium against mycotoxins. Biological Trace Element Research. 2019;**190**(1):273-280

[87] Le G, Yang L, Du H, Hou L, Ge L, Sylia A, et al. Combination of zinc and selenium alleviates ochratoxin A-induced fibrosis via blocking ROS-dependent autophagy in HK-2 cells. Journal of Trace Elements in Medicine and Biology Organ Society Miner Trace Elem GMS. 2022;**69**:126881

[88] Mosallam FM, El-Sayyad GS, Fathy RM, El-Batal AI. Biomolecules-mediated synthesis of selenium nanoparticles using aspergillus oryzae fermented Lupin extract and gamma radiation for hindering the growth of some multidrug-resistant bacteria and pathogenic fungi. Microbial Pathogenesis. 2018;**122**:108-116

[89] Bafghi MH, Nazari R, Darroudi M, Zargar M, Zarrinfar H. The effect of biosynthesized selenium nanoparticles on the expression of CYP51A and HSP90 antifungal resistance genes in aspergillus fumigatus and aspergillus flavus. Biotechnology Progress. 2022;**38**(1):e3206

[90] Parsameher N, Rezaei S, Khodavasiy S, Salari S, Hadizade S, Kord M, et al. Effect of biogenic selenium nanoparticles on ERG11 and CDR1 gene expression in both fluconazole-resistant and -susceptible Candida albicans isolates. Current Medicine Mycology. 2017;**3**(3):16-20

[91] Lotfali E, Toreyhi H, Makhdoomi Sharabiani K, Fattahi A, Soheili A, Ghasemi R, et al. Comparison of antifungal properties of gold, silver, and selenium nanoparticles against amphotericin B-resistant Candida glabrata clinical isolates. Avicenna Journal of Medical Biotechnolog. 2021;**13**(1):47-50

[92] Nile SH, Thombre D, Shelar A, Gosavi K, Sangshetti J, Zhang W, et al. Antifungal properties of biogenic selenium nanoparticles functionalized with nystatin for the inhibition of Candida albicans biofilm formation. Molecular Basel Switzerland. 2023;**28**(4):1836

[93] Nuttall KL. Evaluating selenium poisoning. Annals of Clinical and Laboratory Science. 2006;**36**(4):409-420

[94] Wang Z, Forceville X, Van Antwerpen P, Piagnerelli M, Ahishakiye D, Macours P, et al. A large-bolus injection, but not continuous infusion of sodium selenite improves outcome in peritonitis. Shock Augusta Ga. 2009;**32**(2):140-146

[95] Forceville X, Laviolle B, Annane D, Vitoux D, Bleichner G, Korach JM, et al. Effects of high doses of selenium, as sodium selenite, in septic shock: A placebo-controlled, randomized, double-blind, phase II study. Critical Care London England. 2007;**11**(4):R73

[96] Fraga CG. Relevance, essentiality and toxicity of trace elements in human health. Molecular Aspects of Medicine. 2005;**26**(4-5):235-244

[97] Ikram M, Javed B, Raja NI, Mashwani ZUR. Biomedical potential of plant-based selenium nanoparticles: A comprehensive review on therapeutic and mechanistic aspects. International Journal of Nanomedicine. 2021;**16**:249-268

[98] Wang D, Rensing C, Zheng S. Microbial reduction and resistance to selenium: Mechanisms, applications and prospects. Journal of Hazardous Materials. 2022;**421**:126684

[99] Truong LB, Medina-Cruz D, Mostafavi E, Rabiee N. Selenium nanomaterials to combat antimicrobial resistance. Molecular Basel Switzerland. 2021;**26**(12):3611

[100] Johnson CC, Fordyce FM, Rayman MP. Symposium on "geographical and geological influences on nutrition": Factors controlling the distribution of selenium in the environment and their impact on health and nutrition. The Proceedings of the Nutrition Society. 2010;**69**(1):119-132

[101] Stoffaneller R, Morse NL. A review of dietary selenium intake and selenium status in Europe and the Middle East. Nutrients. 2015;**7**(3):1494-1537

[102] Santos LR, Neves C, Melo M, Soares P. Selenium and Selenoproteins in immune mediated thyroid disorders. Diagnosis Basel Switzerland. 2018;**8**(4):70

[103] Ashton K, Hooper L, Harvey LJ, Hurst R, Casgrain A, Fairweather-Tait SJ. Methods of assessment of selenium status in humans: A systematic review. The American Journal of Clinical Nutrition. 2009;**89**(6):2025S-2039S

Chapter 5

Photic Stress and Rhythmic Physiological Processes: Roles of Selenium as a Chronobiotic

Ayoola Awosika, Mayowa J. Adeniyi, Akhabue K. Okojie and Cynthia Okeke

Abstract

Physiological processes exhibit distinct rhythmic patterns influenced by external cues. External cues such as photic signal play an important role in the synchronization of physiological rhythms. However, excess of or indiscriminate exposure to photic signals exerts profound effects on physiological processes, disrupting normal hormonal secretory rhythms, altering sleep/wakefulness cycle, and impairing reproductive function. Alteration in sleep/wakefulness cycle, impairment in reproductive cycle, and disruption of normal hormonal secretory rhythms characterize risk groups for photic stress such as night workers, trans-meridian travelers, and night-active people. Evidence from primary studies is increasing on the tendency of selenium to reset internal biorhythms by targeting circadian proteins and melatonin. The review highlights the chronobiological roles of selenium.

Keywords: selenium, chronobiotic, photic stress, circadian proteins, melatonin, rhythm, chronobiology

1. Introduction

Virtually all physiological processes including gene expressions exhibit rhythmic patterns. These patterns are influenced by external cues such as light, temperature, metabolic activity, and diet. Indiscriminate exposure to external cues affects the pattern of the rhythms [1–3]. For instance, light is necessary as an external cue to reset circadian pacemakers situated in the suprachiasmatic nucleus; indiscriminate exposure to light and photic stress will affect the functionalities of these pacemakers, causing alteration in the rhythmic pattern of gene expressions with attendant impairments in physiological functions [4, 5]. This may cascade into a raised risk level for a number of medical conditions including cancer, diabetes mellitus, cardiovascular disorders, reproductive derangements, and sleep problems [6]. With continuous proliferation, popularization and utilization of artificial light during nighttime, night workers, trans-meridian travelers, and night-active people tend to be at a higher risk of adverse consequences of circadian misalignment and desynchronization if no precautionary measures are observed.

Besides lifestyle changes and precautionary measures to minimize and mitigate circadian disruptions occasioned by alterations in external cues, most especially light, the roles of nutritional factors cannot be overemphasized [7–9]. Selenium is one of the essential micronutrients for mammals. It is chiefly available in soil and water in variable levels. Its level the plant and animal foods is determined by the soil and water concentration of selenium [10]. It can also be added to food as a supplement. The daily recommended intake of the mineral is 55 micrograms/day for both females and males [11].

As far as its functions are concerned, selenium plays roles as a cofactor for glutathione peroxidase, an enzyme that catalyzes the peroxidation of glutathione to form water. This implies that the mineral is necessary for the regulation of oxidative stress, maintenance of oxidant/antioxidant homeostasis, and prevention of DNA oxidation [12], among others. Second, it acts as a cofactor for iodothyronine deiodinase, an enzyme that converts thyroxin to 3,6,3′-tri-iodothyronine. 3,6,3′-Tri-iodothyronine is an active form of thyroid hormone and far more active than thyroxin. Therefore, the deficiency of selenium may lead to the deficiency of thyroid function, and this can manifest as disorders in all organs where thyroid hormone is needed.

Selenium has also been reported to exhibit the tendency to synchronize biorhythms. This ability is an important corrective measure for desynchronization. A study by Zhang and Zarbi [13] indicated how selenium increased the expression of a circadian protein ‘PER2’. PER2 acts as a negative regulator of circadian rhythm, inhibiting the expression of BMAL1 proteins and CLOCK. Primary studies are available to support the roles of selenium as a therapeutic option for desynchronization. The aim of the work was to highlight the chronobiological roles of selenium.

2. Light pollution and photic stress

The quest for fortune has overwhelmed human affinity for nature and natural mechanisms, one of which is natural light/dark cycle [14]. Nowadays, prolonged exposure to light at night is one of the most common forms of light pollution, an inducer of photic stress [5]. It is characterized by alterations in photoperiod. Conditions associated with light at night include night work and insomnia [4].

The genesis of photic stress can be traced back to the discovery of electric bulb by the renowned American inventor Thomas Alva Edison, who developed a deep vacuum incandescent lamp with a carbon cotton filament [6]. However, the first successful attempt to use electricity for lighting was earlier made by Humphrey Davy in 1801, who discovered the incandescence of an energized conductor [6]. Nowadays, due to rapid electricity proliferation, electric lighting has replaced most traditional lighting sources, making human population virtually independent of natural photoperiod of 12 hour light/12 hour dark cycle. As a matter of fact, over one-third of the world population is estimated to live under light polluted areas [15].

The effects of photic stress are of two types: image-forming effects and photoperiodic effects. While the former are characterized by discomfort and disability glare [16], the latter are characterized by disruption of the circadian rhythm, the internal clock that regulates physiological functions [17].

A major impact of exposure to light at night is the inhibition of melatonin production and shift in the circadian phase [4]. Blue light has been shown to be the most effective in the suppression of melatonin secretion [6]. Light-induced suppression of melatonin is due to reduction in postganglionic noradrenergic neural discharge to pineal glands. Since melatonin rhythm is an efferent mechanism that blends exogenous cycle (light/dark cycle) with endogenous cycle, suppressed nocturnal melatonin secretion represents impairment in synchronization [18].

The desynchronization of the circadian rhythm leads to many clinical conditions. For example, studies have shown the link between exposure to artificial light at night and fatigue [19], reduced work productivity [20], diabetes mellitus [21], many different forms of cancer [20], and derangement in female reproductive functions [22]. In humans, a shift in light/dark cycle characterizing shift work and chronic jetlag suppresses the expression of PER1 and PER2 in the suprachiasmatic nucleus and causes delay in acrophases of the circadian expression of PER1, PER2, BMAL-1, and D-site binding protein (DBP) in the liver [23]. There is a difference between the expression pattern of circadian genes in suprachiasmatic nucleus and peripheral tissues. Yamazaki *et al.* [24] reported that suprachiasmatic nucleus rapidly adjusts to light shifts, but peripheral tissues shift more slowly. For example, PER2 expression in the ovary peaks at light offset delayed by 4–6 hours relative to its expression in the suprachiasmatic nucleus [25]. Also, the duration of light exposure determines whether there will be shifts in the circadian rhythm in both humans and animals [26].

3. Photic stress and rhythmically controlled physiological processes

Biorhythms are periodic variations in physiologic events occurring within a time frame. Important attributes of biorhythms include orderliness, entrainability, self-sustenance, and endogeny [1, 27, 28]. Biorhythms that are completed in less than 24 hours are called ultradian rhythms (example is ultradian LH secretion). It takes more than 24 hours for infradian rhythms to be completed (example is LH surge). Those that are completed in approximately 24 hours are circadian rhythms (example is melatonin secretion).

Circadian rhythms work through a set of expressed proteins known as circadian proteins situated in the suprachiasmatic nucleus in the highest density and other nucleated cells. PER, one of the circadian proteins, interacts with other PER proteins as well as the E-box regulated, clock controlled proteins CRY1 and CRY2 to create a heterodimer, which translocate into the nucleus. At this point, it inhibits CLOCK-BMAL-1 activation [29]. The PER1 mRNA is expressed in all cells as a component of a transcription-translation negative feedback mechanism, which creates a cell autonomous molecular clock. PER1 transcription is controlled by protein interactions and with its 5 E-box and 1 D-box elements in its promoter region. Heterodimer CLOCK-BMAL1 stimulates E-box elements present in the PER1 promoter as well as activates the E-box promoters of other components of the molecular clock such as PER2, CRY1, and CRY2 (**Figure 1**) [5].

Activators include BMAL1 (B); CLOCK (C) and repressors include *period (per)* and *cryptochrome (cry)* and are expressed rhythmically and phosphorylated by Casein kinases (CK) in granulosa cells. Transactivation by BMAL1:CLOCK is indicated by (+); repression of BMAL1:CLOCK activity by PER:CRY is indicated by (−). Arrowheads attached to sine waves indicate rhythmic transcription/translation. Curved arrows indicate nuclear translocation. Abbreviations: arachidonic acid (AA); prostaglandin E2 (PGE2); prostaglandin F2α (PGF2α); phosphorylation (P); Casein kinase 1,2 (CK1,2).

Cyclooxygenase-2 (COX-2), an enzyme involved in prostaglandin synthesis, contains E-box sequences in its promoter region. Studies by Morris and Richard [31]

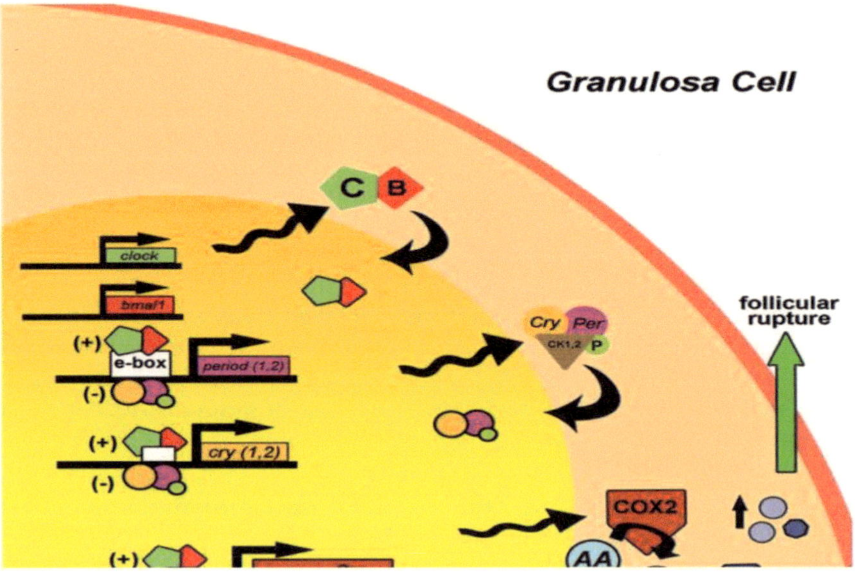

Figure 1.
Molecular mechanism of circadian rhythm in relation to ovulation [30].

and Liu *et al.* [32] showed that CLOCK:BMAL1 heterodimers may activate COX-2 transcription. Circadian rhythms of COX-2 mRNA expression may result in rhythmic buildup of COX-2, which may then result in rhythmic synthesis and accumulation of prostaglandin E2 (PGE2) and prostaglandin F2α (PGF2α). High levels of prostaglandin synthesis, particularly in response to a surge in LH secretion, orchestrate follicular rupture and ovulation.

Hormone secretory pattern and sleep and wakefulness cycle are rhythmic physiological processes. They are influenced by external cues such as light, temperature, and anthropogenic factors, among others. Excess of these cues may abolish these processes. For instance, a study conducted by Attarchi *et al.* [33] on a risk group for light pollution (night shift workers) indicated an increase in FSH levels both in daytime and in nighttime and a decrease in melatonin in daytime and nighttime. FSH secretion is known to peak in the morning and reach nadir level at night, while melatonin is known to peak at around 2.00 am at night and reach nadir during the daytime. The findings of Attarchi *et al.* showed derangement in the normal secretory pattern of FSH and LH. Enormous studies have reported how prolonged exposure to light including light at night affects sleep onset, sleep quality, and sleep duration [4, 5]. Exposure to light before bedtime has been known to delay sleep onset, reduce sleep duration, and impair quality of sleep [34]. Such disruption in sleep/wakefulness cycle increases the risk of individuals acquiring a disease or exacerbates the symptoms of a preexisting condition. Shift work has been associated with an increased risk of mood disorders, depression, cardiovascular disease, endometriosis, and dysmenorrhea as well as an increased incidence and risk of breast cancer [4, 35, 36].

Reproduction involves barrage of rhythmical physiological processes to come by. For instance, at puberty, it is not secretion of gonadotropin-releasing hormone (GnRH) that triggers the episode of changes characterizing the stage but pulsatile secretion of the hormone (occurs every 90 minutes). The circadian rhythms of clock-gene expression noticed in brain areas concerned with reproduction indicate that this neural timing system elicits neuroendocrine events that produce pre-ovulatory luteinizing hormone (LH) surge and ovulation [30]. Works have documented that suprachiasmatic nucleus (SCN) is essential for normal functioning of the hypothalamic pituitary gonadal (HPG) axis [30]. SCN communicates with GnRH neurons through arginine vasopressin (AVP) and vasoactive intestinal peptide (VIP) [25]. The principal afferent pathway to SCN is the photic signal-related retino-hypothalamic pathway. These photic signals are conveyed by light-sensitive retinal ganglionic cells, which do not participate in vision [4, 5], resulting in the control of melatonin production by pineal gland and shift in the circadian phase. Melatonin plays an important role in the photoperiod-induced timing of physiological functions including the cascade of reproductive functions [5, 37].

Excess exposure to light brings about adverse health and reproductive features since circadian clocks are entrained by light duration. For example, shift duty, an employment practice meant to provide service round the clock [38] that is characterized by altered photoperiod and desynchronization of circadian clock, results in health and reproductive problems [5].

Indiscriminate exposure to light has been shown to impair hormonal rhythm, most especially in the hypothalamic hypophyseal ovarian axis, which determines the reproductive cycle and fertility [39]. For instance, continuous illumination was reported to modulate normal nighttime reduction in FSH secretion in women [40]. Other studies indicate that a shift in light/dark cycle by 6 hours caused desynchronization for more than 6 days but requires 6–12 days for clock genes rhythms to completely adjust with different peripheral tissues [24]. Ovarian clock was not fully resynchronized 6 days after exposure to 6 hours shift in light/dark cycle. It took 12 days for full restoration to occur.

Shift workers and trans-meridian travelers tend to have activity, body temperature, and hormonal rhythms that are out of phase with environmental cues [4]. Such disruption may result in endometriosis, dysmenorrhea, as well as an increased incidence and risk of breast cancer [4, 35]. Women working an evening shift, night shift, or irregularly scheduled shifts showed altered menstrual cycle length, increased menstrual pain, and changes in the duration and amount of menstrual bleeding [41]. These symptoms are followed by alterations in patterns of ovarian and hypophyseal hormone secretion, such as an increase in follicular stage length and changes in follicular stimulating hormone (FSH) concentrations [41].

Shift duty is one of the risk factors for photic stress. Female shift workers have been shown to exhibit a higher risk of producing premature or low birth weight babies, spontaneous abortion, and subfecundity [4]. Photopollution has been documented to result in the prolongation of estrous cycle length [15, 42–44], increase in estrous cycle ratio [1, 15, 42, 43], depression in LH, estradiol and progesterone secretions, and increase in estradiol/progesterone ratio [15, 42, 43].

4. Selenium

Selenium is a period IV and group VI element. The major dietary origins of selenium in most countries are plants [10, 45]. Hence, soil selenium concentrations are principal determinants of the minerals in plants and humans [46]. The level of the mineral in the body also depends on state of activity, dialysis, oral contraceptive use, diurnality, pregnancy, and lactation [47, 48], among others. The daily allowance of the mineral is 55 micrograms according to the National Institute of Medicine without gender-related variation.

5. Chronobiotic roles of selenium: Effect on circadian genes

Selenium has been known to be essential for the execution of many physiological functions. As a co-factor for glutathione peroxidase, it is essential for the regulation of oxidative stress. As an antioxidant, glutathione peroxidase helps in the membrane integrity maintenance, prostacyclin production protection, and control of oxidations of macromolecules such as lipids, lipoproteins, and deoxyribonucleic acid (DNA) [49]. As a co-factor for iodothyronine deiodinase, the mineral plays crucial roles in the conversion of tetraiodothyronine (thyroxine) to triiodothyronine, with the latter being an active form of the former [10, 45, 46]. Triiodothyronine is a metabolic hormone. Thus, it exerts its effect on virtually all body tissues. Selenoprotein P is the principal supplier of selenium to tissues [50]. Therefore, free selenium is present in gonads, adrenal gland, thyroid gland, liver, and muscles, among others, whose functions remain sketchy. Selenoprotein P is the main provider of selenium to tissues [50]. Yet low blood and tissue selenium levels have been identified in a number of pathological conditions including HIV infections, cardiomyopathy, and kidney disorder, among others [46, 51].

Another stunning function of selenium is its role in synchronization of circadian clocks. This is predicated by its ability to increase the expression of circadian genes. Synchronization of circadian clocks is essential not only in health but also in copious disease conditions. Since circadian rhythm derangements characterize shift or rotatory work schedule and jetlag and are known as an important risk factor for tumor development (in breast, colon, and prostate), the role of selenium as a chronobiotic cannot be undersized. A study by Hu *et al.* [52] indicated the roles of selenium on circadian gene. L-methyl-selenocysteine was shown to up-regulate BMAL1 in cultured cells and *in vivo* study using mice at the transcription level. As far as the cultured cells were concerned, the authors reported that selenium executed its effects by disrupting TIEG1-induced BMAL1 repression. Conversely, in CLOCK mutant mice deficient in BMAL1, selenium could not orchestrate protection. BMAL1 plays an important role in the positive regulation or activation of circadian rhythm by bringing about the expression of PERIOD genes and CRYPTOCHROME.

Circadian genes control DNA repair mechanisms, and DNA repair mechanisms are normal responses to DNA damage. Zarbl and Fang [53] reported that methylselenocysteine improved PER2 expression in experimentally induced mammary carcinogenesis, thus resulting in the inhibition of mammary tumor development. In an early study, Zhang and Zarbi [13] showed that methylselenocysteine dietary administration at 3 ppm caused time-related and progressive elevation in circadian controlled transcription factor DBP and PER2 gene expression in mammary gland. Conversely, rats placed on standard chow exhibited little or no circadian fluctuation. In *N*-nitroso-*N*-methylurea-induced mammary carcinogenesis, selenium administration reduced circadian controlled transcription factor DBP and PER2 gene expression over time, while no change was noticed in those that were on normal standard chow, but the proteins were more expressed in selenium-treated carcinogenic rats than in untreated carcinogenic rats.

DOI: http://dx.doi.org/10.5772/intechopen.110294

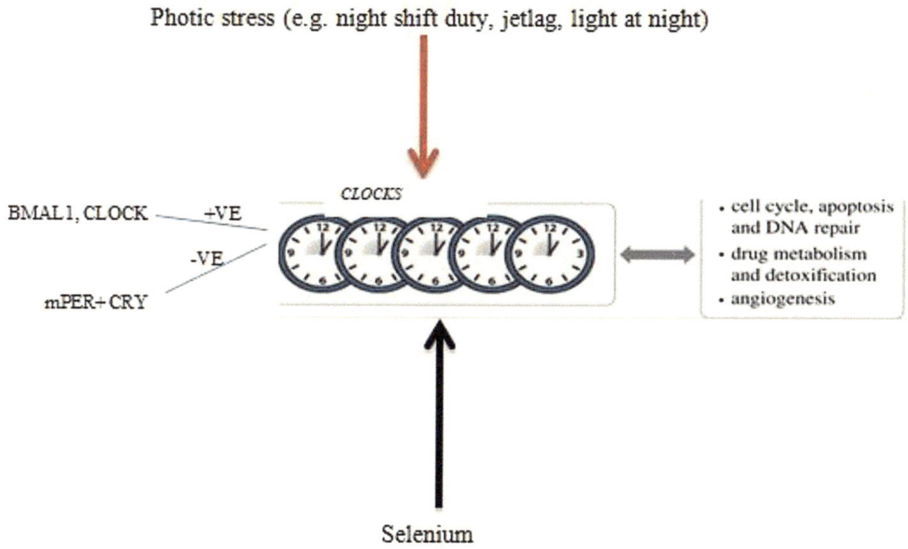

Figure 2.
Effect of selenium and photic stress on circadian clock. Thick black line (stimulation), thick red line (inhibition); +VE (activation), −VE (negative feedback).

DNA methylation, gene expression, and histone protein modification are controlled by circadian rhythms. Xiang *et al.* [54] observed that treatments with selenite reduced DNA methyltransferase mRNA expression and 1 and 3A and protein levels of DNA methytransferase 1 in human prostatic carcinoma cell line (LNCaP cells). The effect of selenium administrations on PER1 expression in normal and desynchronized rats has been reported [44]. In the study, rats were desynchronized through exposure to experimental model of light pollution and photic stress for 1 week and 8 weeks. Dampening of PER1 expression was observed when compared to rats maintained under a natural 12-hour light/12-hour dark cycle. Conversely, administrations of selenium to normal rats for 8 weeks increased the expression of the clock gene. There was also an increase in PER1 expression when selenium was administered for 1 week and 8 weeks to desynchronized rats. The findings of the study suggest the tendency of selenium to resynchronize rats and provide insights into potentials of using selenium as a nutritional alternative for the prevention of diverse adverse alteration induced by excessive exposure to light as occurs in shift duty workers and people who may be exposed to artificial light (**Figure 2**).

6. Chronobiotic roles of selenium: Effect on melatonin synthesis

Another important facet of chronobiology has to do with the regulation of melatonin rhythms. Melatonin is a renowned chronobiotic; it shifts in circadian phase [55], thereby affecting sleep–wake timing, blood pressure regulation, and reproduction [56]. Melatonin synthesis regulation is one of the principal outputs of light-related retino-hypothalamic pathway. During daytime, light rays enter the superior cervical ganglion through the retino-hypothalamic tract and reduce the expression of arylakylamine N acetyl transferase (ANAT), a rate-limiting enzyme that converts serotonin to melatonin. Hence, serotonin, a mood and alertness chemical messenger, becomes high in the day. Reverse occurs in the night. Epinephrine induces the expression of ANAT, raising melatonin level. Melatonin then binds with its receptors in the hypothalamus, retina, and anterior pituitary gland and reduces cAMP. This culminates into reduction in metabolic activities and sleep.

Administration of melatonin to subjects with impaired sleep/wakefulness cycle leads to resynchronization and normalization of the sleep/wakefulness cycle. Any underlying mechanism may include the influence of melatonin on clock gene expression. A study by Adeniyi *et al.* [44] indicated a positive correlation between nocturnal melatonin secretion and ovarian PER1 expression.

Works have shown the influence of selenium on melatonin secretion in living organisms. Adeniyi *et al.* [28] reported that selenium supplementation increased melatonin secretion when compared with rats that were not administered selenium. But when rats were maintained under prolonged dark condition and concomitantly treated with selenium, there was reduction in melatonin secretion. Selenite administered exogenously increased the endogenous secretion of melatonin. This occurs through the control of melatonin synthesis genes such as TDC, T5H, SNAT, and COMT [57]. In a similar pattern, Sun *et al.* [58] reported that selenite at a dose of 96 micrograms/kg increased melatonin synthesis. At 100 micrograms/kg and 150 micrograms/kg of selenium administrations, there was an increase in melatonin secretion in rats. In rats that were exposed to excess light, selenium administration at 150 micrograms/kg increased melatonin secretion after 1 week and 8 weeks of treatments [44].

7. Discussion

Suprachiasmatic nucleus of the hypothalamus is known as a master clock as it contains the largest amounts of circadian proteins PERIODS, CRYPTOCHROME, BMAL1, and CLOCK [25]. These proteins are also present in peripheral tissues in the body, where they regulate the timing and oscillation of gene expressions and biological events. Suprachiasmatic nucleus receives input signals through many pathways, but the principal is the light-mediating retino-hypothalamic tract, which regulates melatonin secretion and rhythmic proteins and synchronizes the body's endogenous rhythms with external rhythms [30].

Night workers, trans-meridian travelers, and night active people are at a risk of desynchronization, a mismatch between external rhythms, especially light/dark cycle and endogenous rhythms. This mismatch also implies alteration in gene expressions and protein synthesis and variations in physiological processes, thereby aggravating the likelihood of sleep problems, endocrine disorders, reproductive derangements, and cancers [3, 6, 15, 34, 42, 43]. Specifically, breast cancer development likelihood has been reported in observers of night duty [4, 6]. In view of the necessity of night work in a teeming and ever-demanding world, the need for diverse palliatives is inevitable.

Selenium is a possible nutritional palliative for chronobiological problems in view of its ability to increase circadian genes and melatonin. Circadian proteins and melatonin determine the characteristics of rhythms and control gene expressions in nearly all body tissues. Insights into the possibility of selenium retarding tumor development stemmed from an observation that experimental rats administered selenium-enriched garlic exhibited declined cancer development [59, 60]. Although more primary studies are needed to authenticate the doses of different forms of selenium required to achieve this chronobiological effects not only in experimental animals but also in humans, the increase in PER2 expression by mammary tissue by selenium as reported by Zhang and Zarbi [13] and an increase in the expression of the clock gene in selenium-treated carcinogenic rats when compared with untreated N nitroso N methylurea-induced mammary carcinogenesis indicate that PER2 is a target of selenium. In a similar development, selenium administrations at 100 micrograms/kg and 150 micrograms/kg increased the PER1 expression in the ovaries of female rats exposed to photic stress via prolonged lighting period [15, 42, 43].

Melatonin has been used to treat sleep disorders for years as a chronobiotic. That selenium, a naturally occurring element, present in plant and animal foods can increase melatonin is quite remarkable and may reduce abusive use of melatonin for sleep induction. Evidence of its tendency to alleviate and mitigate circadian disruptions and reproductive derangements in animal studies [28, 44] is also thrilling. However, more studies are required to prove the level of safety associated with the use and prolonged use of selenium in human beings.

8. Conclusion

The review has highlighted biorhythmic effects of photic stress and the chronobiological roles of selenium.

Author details

Ayoola Awosika[1], Mayowa J. Adeniyi[2]*, Akhabue K. Okojie[3] and Cynthia Okeke[4]

1 University of Illinois, College of Medicine, Chicago, USA

2 Departments of Physiology, Federal University of Health Sciences Otukpo, Nigeria

3 Department of Physiology, Baze University, Abuja, Nigeria

4 Department of Physiology, Igbinedion University Okada, Nigeria

*Address all correspondence to: 7jimade@gmail.com

References

[1] Adeniyi MJ, Agoreyo FO. Estrous cycle ratio as a reproductive index in the rats. American Journal of Biomedical Science & Research. 2019;**4**(2):100-103

[2] Adeniyi M. Impacts of environmental stressors on autonomic nervous system. In: Aslanidis T, editor. M.Sc. NourisAutonomic Nervous System - Special Interest Topics [Working Title]. London, UK, London, UK: IntechOpen; 2022. DOI: 10.5772/intechopen.101842

[3] Adeniyi M, Olaniyan O, Fabunmi O, Okojie A, Ogunlade A, Ajayi O. Modulatory role of pre-exercise water ingestion on metabolic, cardiovascular and autonomic responses to prolonged exercise in young mildly active male. International Journal of Biomedical Sciences. 2022;**18**(2):1-9

[4] Mahoney MM. Shift work, jet lag, and female reproduction. International Journal of Endocrinology. 2010;**2010**:813764. DOI: 10.1155/2010/813764. Epub 2010 Mar 8. PMID: 20224815; PMCID: PMC2834958

[5] Lowden A, Moreno C, Holmback U, Lennernas M, Tucker P. Eating and shift work–effects on habits, metabolism and performance. Scandinavian Journal of Work, Environment & Health. 2010;**36**:150-162

[6] Haim A, Portnov BA. Light Pollution as a New Risk Factor for Human Breast and Prostate Cancers. Dordrecht Heidelberg New York London: Springer; 2013 ISBN 978-94-007-6219-0

[7] Adeniyi MJ, Ige SF. Therapeutic effect of crude extract of *Garcinia kola* (bitter Kola) on acetic acid induced colitis in male rats. International Journal of Healthcare Sciences. 2016;**4**(2):1141-1149

[8] Adeniyi MJ, Ige SF, Adeyemi WJ, Oni TJ, Ajayi PO, Odelola SO, et al. Changes in body temperature and intestinal antioxidant enzymes in healthy and colitis male rats: Roles of *Garcinia kola* (bitter Kola). International Journal of Physiology. 2016;**4**(2):36-41

[9] Ige SF, Adeniyi MJ, Iyalla GO. *Allium cepa* mitigates aluminum chloride-induced hepatotoxicity in male Wistar rats. Journal of Biomedical Science. 2017;**6**(4):27

[10] Rayman MP. Selenium and human health. Lancet. 2012;**379**:1256-1268

[11] Monsen ER. Dietary reference intakes for the antioxidant nutrients: Vitamin C, vitamin E, selenium, and carotenoids. Journal of the American Dietetic Association. 2000;**100**(6):637-640

[12] Mistry HD, Pipkin FB, Redman CW, Poston L. Selenium in reproductive health. American Journal of Obstetrics & Gynecology. Jan 2012;**206**(1):21-30. DOI: 10.1016/j.ajog.2011.07.034. Epub 2011 Jul 29. PMID: 21963101

[13] Zhang X, Zarbl H. Chemopreventive doses of methylselenocysteine alter circadian rhythm in rat mammary tissue. Cancer Prevention Research (Philadelphia, Pa.). 2008;**1**:119-127

[14] Stevens RG, Brainard GC, Blask DE, Lockley SW, Motta ME. Adverse health effects of nighttime lighting: Comments on American Medical Association policy statement. American Journal of Preventive Medicine. 2013;**45**(3):343-346

[15] Adeniyi MJ, Agoreyo FO. Duration-related modulation of body temperature rhythm and reproductive cycle in rats by photoperiodic perturbation. DRJHP. 2020;**8**(1):1-6

[16] Kent MG. Temporal Effects in Glare Response. 'Ph.D. Thesis. Nottingham, UK: University of Nottingham; 2016

[17] Touitou Y, Reinberg A, Touitou D. Association between light at night, melatonin secretion, sleep deprivation, and the internal clock: Health impacts and mechanisms of circadian disruption. Life Sciences. 2017;**193**:94-106

[18] Moore-Ede M, Platika D. Health risks of light at night the good, the bad, and the time of day of bioactive blue light. A Circadian White Paper. 2016;**02180**(781):439-6333

[19] Weinert D, Waterhouse J. Interpreting circadian rhythms. In: Biological Timekeeping: Clocks, Rhythms and Behaviour. New Delhi, India: Springer; 2017. pp. 23-45

[20] Lim HS, Ngarambe J, Kim JT, Kim G. The reality of light pollution: A field survey for the determination of lighting environmental management zones in South Korea. Sustainability. 2018;**10**:374

[21] Hu C, Jia W. Linking MTNR1B variants to diabetes: The role of circadian rhythms. Diabetes. 2016;**65**:1490-1492

[22] Yorshinaka K, Yamaguchi A, Matsumura R, Node K, Tokuda I, Akashi M. Effect of different light-dark schedules on estrous cycle in mice and implications for mitigating the adverse impact of night work. Genes to Cells. 2017;**22**(10):876-884

[23] Khan S, Duan P, Yao L, Hou H. Shiftwork-mediated disruptions of circadian rhythms and sleep homeostasis cause serious health problems. International Journal of Genomics. 2018;**2018**:11

[24] Yamazaki S, Numano R, Abe M, Hida A, Takahashi R, Ueda M, et al. Resetting central and peripheral circadian oscillators in transgenic rats. Science. 2000;**288**(5466):682-510

[25] Toffol E, Heikinheimo O, Partonen T. Biological rhythms and fertility: The hypothalamus–pituitary–ovary axis. ChronoPhysiology and Therapy. 2016;**6**:15-27

[26] Dewan K, Benloucif S, Reid K, Wolfe LF, Zee PC. Light-induced changes of the circadian clock of humans: Increasing duration is more effective than increasing light intensity sleep. Sleep. 2011;**34**(5):593-599

[27] Awosika A, Adeniyi M. Impact of post-exercise Orthostasis on EEG amplitudes in age-matched students: Role of gender. Archieves in Neurology & Neuroscience. 2022;**13**:1-4

[28] Adeniyi MJ, Agoreyo FO, Abayomi AA. Pineal and Hypophyseal responses to selenium treatments in light deprived adult female Wistar rats. IBBJ. 2019;**6**:1. Available from: http://ibbj.org/article-1-243-en.html

[29] Ko CH, Takahashi JS. Molecular components of the mammalian circadian clock. Human Molecular Genetics. 2006;**15**(2):R271-R277

[30] Sellix MT, Menaker M. Circadian clocks in the ovary. Trends in Endocrinology and Metabolism. 2011;**21**(10):628-636

[31] Morris JK, Richards JS. An E-box region within the prostaglandin endoperoxide synthase-2 (PGS-2) promoter is required for transcription in rat ovarian granulosa cells. The Journal of Biological Chemistry. 1996;**271**:16633-16643

[32] Liu J, Antaya M, Boerboom D, Lussier JG, Silversides DW, Sirois J. The

delayed activation of the prostaglandin G/H synthase-2 promoter in bovine granulosa cells is associated with down-regulation of truncated upstream stimulatory factor-2. The Journal of Biological Chemistry. 1999;**274**:35037-35045

[33] Attarchi M, Darkhi H, Kashanian M, Khodarahmian M, Dolati M, Ghaffari M, et al. Characteristics of menstrual cycle in shift workers. Global Journal of Health Science. 28 Feb 2013;**5**(3):163-172. DOI: 10.5539/gjhs.v5n3p163. PMID: 23618486; PMCID: PMC4776814

[34] Blume C, Garbazza C, Spitschan M. Effects of light on human circadian rhythms, sleep and mood. Somnologie (Berl). 2019;**23**(3):147-156. DOI: 10.1007/s11818-019-00215-x Epub 2019 Aug 20

[35] Turek FW. From circadian rhythms to clock genes in depression. International Clinical Psychopharmacology. 2007;**22**(2):S1-S8

[36] Adeniyi MJ, Oni TJ. Water contamination in Nigeria and body defense issues. Research Chronicler. 2016;**3**(3):11-21

[37] Olayaki LA, Soladoye AO, Salman TM, Joraiah B. Effects of photoperiod on testicular functions in male Sprague-Dawley rats. Nigerian Journal of Physiological Sciences. 2008;**23**(1-2):27-30

[38] Costa G. Shift work and health: Current Prob-lems and preventive actions. Safety Health Work. 2010;**1**:112-123

[39] Wang Y, Gu F, Deng M, Guo L, Lu C, Zhou C, et al. Rotating shift work and menstrual characteristics in a cohort of Chinese nurses. BMC Women's Health. 2016;**16**:24

[40] Davis S, Mirick DK, Chen C, Stanczyk FZ. Night shift work and hormone levels in women. Cancer Epidemiology, Biomarkers & Prevention. 2012;**21**(4):609-618

[41] Chung F-F, Yao C-CC, Wan G-H. The associations between menstrual function and life style/working conditions among nurses in Taiwan. Journal of Occupational Health. 2005;**47**(2):149-156

[42] Adeniyi MJ, Fabunmi O, Okojie AK, Olorunnisola OL, Odetola AO, Olaniyan OT, et al. Impact of night study frequency on sleep pattern, anthropometrical indices and peripheral oxygen saturation in age-matched Nigerian female students prior to semester examination. International Journal of Biomedical Sciences. 2020a;**16**(3):37-42

[43] Adeniyi MJ, Agoreyo FO, Olorunnisola OL, Olaniyan OT, Seriki SA, Ozolua PO, et al. Photo-pollution disrupts reproductive homeostasis in female rats: The duration-dependent role of selenium administrations. The Chinese Journal of Physiology. 2020b;**63**:235-243

[44] Adeniyi MJ, Agoreyo FO, Ige SF, Fabunmi OA, Ozolua OP, Biliaminu SA. Physiological implications of prolonged selenium administrations in Normal and desynchronized adult female rats. Journal of Krishna Institute of Medical Sciences (JKIMSU). 2021;**10**(4):21-36

[45] Rayman MP. Selenium and human health. Lancet. 2008;**379**:1256-1268

[46] Adeniyi MJ, Agoreyo FO. Nigeria and the selenium micronutrient: A review. Annals of Medical and Health Sciences Research. 2018;**8**:5-11

[47] Adeniyi MJ, Agoreyo FO. Diurnal effect of selenium supplementations on adult female Wistar rats made hypothyroid by Methimazole. Biomedical Journal of Scientific & Techmical Research. 2017;**1**(2):283-285

[48] Agoreyo FO, Adeniyi MJ. Pattern of estrous cycle and ovarian Antiperoxidative activity in light deprived Sprague-Dawley rats treated with sodium selenite. Journal of Medicinal Research and Biological Studies. 2018;**1**(1):103

[49] Brigelius-Flohe R, Banning A, Schnurr K. Selenium-dependent enzymes in endothelial cell function. Antioxidants & Redox Signaling. 2003;**5**:205-215

[50] Hurst R, Armah CN, Dainty JR, Hart DJ, Teucher B, Goldson AJ, et al. Establishing optimal selenium status: Results of a randomized, double-blind, placebo-controlled trial. The American Journal of Clinical Nutrition. 2010;**91**:923-931

[51] Anyabolu HC, Adejuyigbe EA, Adeodu OO. Serum micronutrient status of Haart-Naïve, HIV infected children in south West Nigeria, a case controlled study. AID Research and Treatment. 2014;**2014**:351043. DOI: 10.1155/2014/351043. Epub 2014 Aug 11. PMID: 25180086; PMCID: PMC4144154

[52] Hu Y, Spengler ML, Kuropatwinski KK, Comas-Soberats M, Jackson M, Chernov MV, et al. Selenium is a modulator of circadian clock that protects mice from the toxicity of a chemotherapeutic drug via upregulation of the core clock protein, BMAL1. Oncotarget. 2011;**2**(12):1279-1290

[53] Zarbl H, Fang M. Dietary Methylselenocysteine and epigenetic regulation of circadian gene expression. In: Patel V, Preedy V, editors. Handbook of Nutrition, Diet, and Epigenetics. Cham: Springer; 2017. DOI: 10.1007/978-3-319-31143-2_63-1

[54] Xiang N, Zhao R, Song G, Zhong W. Selenite reactivates silenced genes by modifying DNA methylation and histones in prostate cancer cells. Carcinogenesis. 2008;**29**(11):2175-2181. DOI: 10.1093/carcin/bgn179 Epub 2008 Aug 1

[55] Hardeland R, Pandi-Perumal SR, Cardinali DP. Melatonin. The International Journal of Biochemistry & Cell Biology. 2006;**38**(3):313-316

[56] Altun A, Ugur-Altun B. Melatonin: Therapeutic and clinical utilization. International Journal of Clinical Practice. 2007;**61**(5):835-845

[57] Yang N, Sun K, Wang X, Wang K, Kong X, Gao J, et al. Melatonin participates in selenium-enhanced cold tolerance of cucumber seedlings. Frontiers in Plant Science. 2021;**12**:786043. DOI: 10.3389/fpls.2021.786043

[58] Sun X, Luo Y, Han G, et al. Effects of low-dose selenium on melatonin synthesis in sweet cherry. Journal of Soil Science and Plant Nutrition. 2021;**21**:3309-3319

[59] Ip C, Lisk DJ. Efficacy of cancer prevention by high-selenium garlic is primarily dependent on the action of selenium. Carcinogenesis. 1995;**16**(11):2649-2652

[60] Ip C, Zhu Z, Thompson HJ, Lisk D, Ganther HE. Chemoprevention of mammary cancer with Se-allylselenocysteine and other selenoamino acids in the rat. Anticancer Research. 1999;**19**(4B):2875-2880

Chapter 6

Replacement Selenium Therapy in Acute Cerebral Damage

Irina Alexandrovna Savvina,
Hasaybat Salimbekovna Nucalova, Anna Olegovna Petrova,
Kristina M. Bykova and Irina Varlamovna Tkebuchava

Abstract

The current literature covers the role of selenium in metabolic processes and the importance of correcting its level in various diseases and critical conditions, including acute cerebral damage due to severe traumatic brain injury (TBI) and sepsis-associated encephalopathy (SAE). Numerous experimental animal studies have demonstrated that selenium has protective properties and blocks the mechanisms of apoptosis, and is involved in maintaining the functional activity of neurons and inhibits astrogliosis. The study of the selenium content in the blood of patients with acute cerebral damage due to severe TBI and sepsis with verified SAE, and the development of schemes of replacement selenium therapy will improve outcomes, both in increasing survival and in reducing the resuscitation bed-day and the number of neurological deficits in the future.

Keywords: replacement selenium therapy, acute cerebral damage, severe traumatic brain injury, sepsis-associated encephalopathy, selenium

1. Introduction

Selenium is an essential trace element in the human body. Normally, the concentration of selenium in blood plasma is 100–200 mcg/l. The most studied function of selenium is the regulation of antioxidant processes in all organs and tissues, primarily in the central nervous system [1]. Selenium deficiency leads to an imbalance of the lipid peroxidation/antioxidant system, which is a constant component of any pathological process [2, 3]. Selenium deficiency provokes structural changes in the membranes of microsomes, and damage to the organoid membranes of cells of almost all tissues and it is accompanied by a change in the activity of 5-nucleotidase, creatine phosphokinase, LDH, b- hydroxybutyrate, AST, ALT, aldolase, Na, and K-ATPase [4, 5].

Experimental animal studies have demonstrated that selenium has protective properties and blocks the mechanisms of apoptosis-cell death, and participates in maintaining the functional activity of neurons and in inhibiting astrogliosis in the acute cerebral injury of various etiologies [6–10].

It is known that selenium-dependent proteins—the family of glutathione peroxidases (GPX1-6), as well as selenoproteins P, W, T, M, etc. play a key role in the processes of inhibition of free radical oxidation chain reactions [4, 5].

The function of glutathione peroxidases is to maintain stable intracellular concentrations of reduced glutathione. Cytosolic glutathione peroxidase (GPX1) plays a major protective role in the development of oxidative stress. GPX1 activity is more dependent on selenium content compared to other enzymes, and therefore GPX1 activity in erythrocytes is a simple and sensitive indicator of the selenium status of the organism [11]. Intracellular and tissue levels of GPX1 also affect the activity of apoptotic pathways and phosphorylation of protein kinases [12].

Selenoprotein P, being the main extracellular source of selenium, performs the function of selenium transport to various tissues, mainly to the brain [13, 14], as well as antioxidant functions [15, 16], normally amounts to 6–7 micrograms of selenium/dl plasma.

Thus, selenium is a very important micronutrient for adequate function of the brain. The role of selenium is to protect against oxidative stress and other damaging factors in the central nervous system [17], maintaining the balance of neurotransmission and inflammation control [18].

2. Selenium metabolism in critical conditions

The study of the dynamics of antioxidant systems and lipoperoxidation processes made it possible to clarify the basic pathophysiological mechanisms underlying the development of critical conditions [19–21]. Activation of the processes of lipid peroxidation and oxidative modification of plasma proteins, which leads to a violation of the structural and functional integrity of membranes, inactivation of protein enzymes, and impaired synthesis of nucleic acids and protein, is a universal damaging mechanism in severe trauma and critical conditions of any genesis [22, 23]. Ischemic-reperfusion injury often accompanies severe forms of systemic inflammatory reactions, exacerbating the harmful effects of free radicals, and leading to an imbalance between oxidation processes and antioxidants. This situation has already been described in patients with sepsis and non-septic forms of systemic inflammatory syndromes in which there is a significant increase in the production of free radicals, especially superoxide anions [24].

In critical conditions, there is an increasing consumption of selenium and insufficient intake of it into the body from the outside, which leads to a deficiency of selenium in the body and makes it defenseless when exposed to oxygen free radicals and the cascade of reactions caused by their activation [25–27].

Selenium exhibits significant antioxidant activity, preventing changes in cell membranes, participates in respiratory chain reactions, in the pentose phosphate cycle, in the citric acid cycle, and lipid peroxidation [28]. Selenium activates protein synthesis, participates in antihistamine and antiallergic mechanisms, and normalizes the metabolism of proteins and nucleic acids [29].

Pathogenetic substantiation for the use of selenium in the intensive care complex in critical conditions, according to a number of authors [30–34], consists in the following mechanisms of action:

- Suppression of endothelial adhesion and protection of the endothelium from damage by oxygen radicals;

- Reduced production of pro-inflammatory cytokines;
- Suppression of hyperactivity of the nuclear factor NF-kB;
- Decrease in the activity of the complement system;
- Maintaining the utilization of peroxides;
- Stabilization of glucocorticoid receptors;
- Stimulation of the insulin signaling cascade due to an insulin-like effect that improves glucose control.

The above mechanisms of action contribute to the prevention of microcirculatory-mitochondrial dysfunction as a universal link of multiple organ failure [35–37].

Selenium plays an important role in the functioning of the immune system. Thus, in conditions of selenium deficiency, the processes of antigen-dependent lymphocyte proliferation, neutrophil chemotaxis are disrupted, and the level of IgA, IgG, and IgM decreases [24, 29].

A relationship was established between the low concentration of selenium in the blood serum and the severity of the condition of patients, and the level of mortality, which served as the basis for the early inclusion of selenium in the intensive care regimen for critical conditions [38–40]. The introduction of sodium pentahydrate selenite ensures normalization of plasma selenium concentration in the next 24 hours, leads to improved functioning immunocompetent cells (increased phagocyte activity, T-killer activity, immunoglobulin synthesis, etc.), contributes to improving clinical outcomes and significantly reducing patient mortality [41, 42]. Appointment of sodium selenite in patients in critical condition with infectious systemic inflammatory response ensures normalization of plasma selenium concentration in the next 24 hours. Numerous studies have shown that among patients in critical condition and suffering from sepsis, among those who underwent correction of selenium deficiency, mortality was significantly lower than in patients who did not receive selenium preparations [43–51]. The combination of selenoprotein P for endothelial protection and the cytotoxic effects of Na2SeO3 against hyperactivated leukocytes may be a promising intervention for early sepsis [52]. Copper-selenium nanoclusters may be an efficient strategy to cure sepsis by *in situ* sulfurization of endogenous H2S, triggering ROS eruptions and photothermal therapy [53].

According to the experts of the Cochrane Collaboration [22], concerning studies on selenium exchange in critically ill patients based on an analysis of seven randomized clinical trials, the quality of the studies was not good enough, the availability of outcome data was often limited, and studies examining the effects of selenium replacement therapy were insufficient in size of the study population. In addition, the main problem of these studies was related to the heterogeneity of the studied patient population, as a result of which the results are presented in the form of random effects. Most of the analyzed papers were statistically insignificant. Based on all of the above, the Cochrane Collaboration experts concluded that there is insufficient evidence of the effectiveness of selenium therapy at the present time in relation to the duration of ventilation, bed-day in intensive care, general hospital bed-day or quality of life after a critical condition, to recommend it for use in patients in critical condition. Meanwhile, some authors believe that the inclusion of selenium-containing drugs in the intensive care complex opens up new horizons in the treatment of critical conditions [54]. Note also that in a systematic review by Berger et al. [28], Shenkin [55] provides data on the feasibility of short courses of intravenous use of selenium in patients in critical condition (burns, serious injuries, sepsis, and stroke).

Since 2009, selenium has been included in the ESPEN recommendations as a pharmacological module (Grade C) [56], since 2010—in the national guidelines for the treatment of sepsis in Germany (Grade C) [57, 58].

Among the patients in critical condition, patients with sepsis and polytrauma, including TBI, require the most attention. The role of selenium in the regulation of inflammatory response and gene transcription mechanisms in patients with polytrauma is discussed by a number of authors [59]. Most patients who are in a prolonged unconscious state suffer sepsis at different periods of their disease against the background of low plasma selenium levels [60]. At the same time, the constant administration of various groups of antibacterial drugs often does not affect the frequency of septic complications development and leads only to the formation of polyresistant flora.

In one study carried out by Chelkeba et al. [61], the antioxidant effect of selenium was researched in 54 patients under critical condition due to severe sepsis and septic shock, or mechanically ventilated for more than 48 hours [61]. Twenty-nine patients (1st group) received 2000 μg of sodium selenite in 100 ml of saline solution within the first 6 hours of sepsis diagnosis, followed by 1500 μg of sodium selenite in 250 ml of saline solution for 12 hours continuously for 14 days, had mortality rates lower (31%) then 25 patients (2nd group) with intensive standard treatment without selenium (40%). Also, it was found a significant increase in GPx-3 levels, which causes a blocking action of the inflammatory cytokines [61].

Another clinical study by G. Landesberg with colleagues [62] showed a negative correlation between pro-inflammatory cytokines and the severity of sepsis and myocardial dysfunction assuming that selenium has no effect in septic patients since this nutrient did not present any long-term effect on the pro-inflammatory cytokines plasma concentration [62].

Kieliszek and Lipinski [63] demonstrated that sodium selenite can oxidize thiol groups in disulfide isomerase proteins of the SARS CoV-2 virus, thus preventing the COVID-19 virus from penetrating the membrane of healthy cells of its possible hosts. Such hypotheses can be considered about selenium since this nutrient is of great importance for inflammatory diseases [63].

The study by Mahmoodpoor et al. [64] did not indicate the presence of adverse events related to the high dose of intravenous sodium selenite and aspects of toxicity from its administration [64].

In one meta-analysis selenium supplementation for severe trauma patients was examined. The current evidence supports that selenium administration decreases the mortality rate and ICU and hospital stays for patients who have sustained major trauma. Selenium supplementation was not associated with infectious complications after major trauma [65]. Selenium administration shows no substantial influence on the 28-day mortality, length of ICU stay, duration of vasopressor therapy, incidence of acute renal failure, and serious adverse events for septic patients [66].

Some multiple-center trials confirm the efficacy of high-dose sodium selenite supplementation in patients with severe sepsis and septic shock to reduce 28-day mortality [67].

However, in Valenta et al. [68] study, it was shown that the 28-day mortality is not decreased after selenium administration in septic patients and in critically ill patients [68].

3. Selenium homeostasis in the brain

Insufficient selenium supply and lack of selenoprotein function have been linked to multiple brain disorders, including neurodegenerative diseases, which have been thoroughly discussed in previous reviews [8, 10]. Conversely, selenium has been suggested as a potential therapeutic agent in the treatment of Alzheimer's disease [11], multiple sclerosis [12], and stroke [13, 69–72].

Great importance is attached to the provision of the body with selenium in the occurrence of neurodegenerative diseases (Alzheimer's disease, Parkinson's disease) [69, 73]. The largest and most well-organized study [74], conducted in 2003–2005 in two provinces of China and included 2000 people, showed that low selenium content in nails directly correlates with a decrease in intelligence in people over 65 years of age ($p < 0.0087$). In this regard, selenium preparations are considered promising in the prevention and treatment of Alzheimer's type dementia. In addition, Thiel and Fowkes [75] showed that the use of an antioxidant complex prevents the development of dementia in children with Down's disease (this population represents the largest cohort with an increased risk of dementia due to overexpression of the superoxide dismutase gene) [75].

Another important potential use of selenium is for Parkinson's disease [76]. It is proved that there is a significant increase in the disease prooxidant processes, and the activity of glutathione reductase and other antioxidant enzymes increases compensatorily [77]. At the same time, a study by Kim et al. [78, 79] showed that the use of selenium significantly weakened the phenomena of oxidative stress caused by methamphetamine in nigrostriatal neurons, thus preventing the development of experimental parkinsonism [78, 79]. Note, however, that the concentration of selenium in the cerebrospinal fluid is increased in all patients with Parkinson's disease.

Perhaps this reflects the increased utilization of selenium under conditions of severe oxidative stress in these patients [80]. Recent studies suggest a significant role of selenium and the enzyme glutathione peroxidase in the pathogenesis of epilepsy [81, 82]. Decreased activity of Se-BP1 (selenium-binding protein 1) pathognomonic for schizophrenia, with exacerbation it decreases to critical figures, and with replenishment, there is an improvement in the condition [83].

An important role is played by the change in the antioxidant status in ischemic stroke. In the study of Zimmermann et al. [84], it was shown that on the first day after a stroke, a significant decrease in selenium levels ($p < 0.01$) was observed against the background of increased glutathione peroxidase activity ($p < 0.01$) [84]. Numerous experimental studies [85, 86] demonstrated distinct neuroprotective properties of selenium in conditions of cerebral ischemia. Ansari et al. [85] demonstrated the neuroprotective effect of different doses of selenium (from 0.05 to 0.2 mg/kg) on models of occlusion of the middle cerebral artery [85]. A study by Yousuf et al. [87] showed that the use of selenium in the form of sodium selenite (0.1 mg/kg) led to a significant recovery of ATP levels in the neurons of rats subjected to cerebral ischemia ($p < 0.05–0.001$) [87]. In addition, there was a decrease in the area of edema and microglia infiltration.

Wray J. R. et al. [88] and Perez A. with colleagues [89] demonstrated the glucocorticoids influence on the selenoproteins regulation [88] and the metabolic effects of glucocorticoids, which include over-eating and excess weight gain [89].

The neuroprotective effect of selenium as a result of selenium replacement therapy in patients with neurological deficiency after subarachnoid hemorrhage of aneurismal etiology was noted by Japanese colleagues [90]. Japanese authors also described

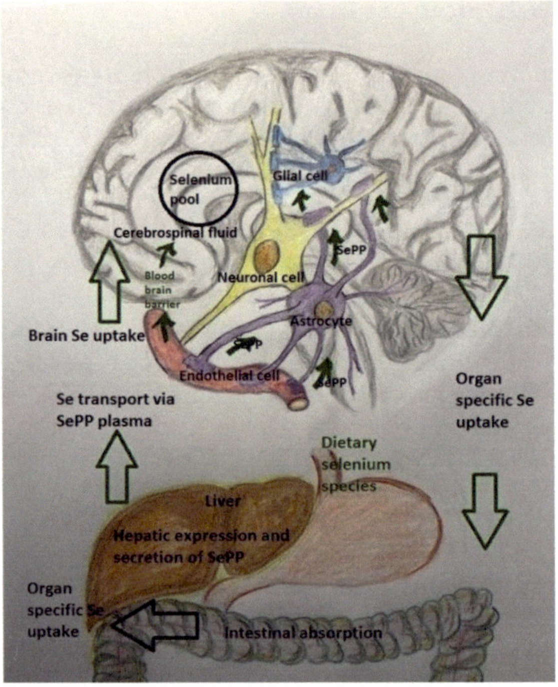

Figure 1.
Scheme of the main physiological processes involved in the mechanism of specific selenium uptake by the brain [92].

the positive effect of the inclusion of ebselen in the complex therapy of ischemic stroke [91]. It should be noted that the selenium-containing drug ebselen is currently undergoing the registration procedure for applications for stroke and subarachnoid hemorrhage in Japan.

The proposed scheme of the main physiological processes involved in the specific mechanisms of selenium uptake by the brain is shown in **Figure 1** [92].

In the experimental study by Xu L. with colleagues [93] was shown that plasma selenium levels were lower in the Chronic Unpredictable Mild Stress (CUMS)-sensitive group of rats [93]. It is important that an epidemiological study correlated low selenium intake with an increased susceptibility for developing the major depressive disorder in humans [94].

4. The role of selenium in preventing apoptosis and cerebral damage (according to the results of experimental studies)

In the experimental works of R.F. Burk et al. [95, 96], it was shown that the introduction of sodium selenite leads to a significant increase in the content of selenoprotein P in the brain (compared with other tissues), and in conditions of selenium deficiency, the brain's uptake of selenoprotein P increases by five times; at the same time, low-molecular selenium compounds are not utilized by the brain [95, 96]. Moreover, the research of P. R. Hoffmann et al. [97] showed that genetic deficiency of selenoprotein P in transgenic mice leads to a decrease in the expression of other selenoproteins in the brain; presumably, this is due to the mechanism of selenoprotein biosynthesis: in conditions of cellular selenium deficiency, the UGA codon encoding selenocysteine begins to play the role of a stop codon, and the synthesis of selenium protein is interrupted [97].

In experiments on rats in the early period of TBI (after 6 hours and 24 hours), there was a sharp decrease in the level of selenium and vitamin E in the blood of animals [7]. The reason for the decrease was oxidative stress and a high level of selenium consumption. Therefore, according to the authors of the study, it is necessary to restore selenium levels to normal values preceding TBI.

In an experimental model of cerebral ischemia/reperfusion in rats created by occlusion of the right carotid artery for 45 minutes, animals were treated with ginkgo biloba (50 mg/kg/day intraperitoneally) and selenium (0.625 mg/kg intraperitoneally) for 14 days after occlusion [98]. The activity of superoxide dismutase and glutathione peroxidase enzymes was measured in hippocampal tissue in 25 animals. An immunohistochemical study was performed with electron and light microscopy. According to the results of the study, the authors concluded that through the suppression of oxidative stress processes, a significant effect of neuroprotection in ischemia/reperfusion is realized with the combined use of ginkgo, selenium, and their combination [98]. Thus, data presented in the study by G. Erbil et al. [98], demonstrate that selenium treatment after ischemic/reperfusion injury improves the activity of antioxidant enzymes, prevents neuronal damage and moderate reactive gliosis caused by this kind of damage in the hippocampus in rats [98].

The inclusion of selenium as monotherapy or in combination with ginkgo significantly reduces brain tissue damage in this experimental model. Casaril A. M. with colleagues [99] showed that 3-((4-chlorophenyl)selanyl)-1-methyl-1H-indole (CMI) can prevent acute stress-induced depressive-like behavior in mice [99]. Also, CMI induces antinociceptive effects in mice by modulating serotonergic activity [100] and can reverse the depressive-like phenotype caused by lipopolysaccharide injection [101].

It is obvious that the results obtained *in vitro* and *in vivo* experiments on rats demonstrate that selenium has a protective effect in ischemic/reperfusion injury in many tissues, including neuronal [102–104].

Oxidative stress, which is a universal pathophysiological mechanism in polytrauma, combined trauma and TBI, leads to the development of reactive gliosis in TBI. Damage to the astroglia may be a significant contribution to the formation of neuronal damage. It is well known that ischemia/reperfusion induces neuronal damage through several pathophysiological mechanisms, including intracellular Ca^{++} movement and free radical production, which ultimately triggers apoptosis. In the body, selenium protects cells from free radicals and peroxidase activity caused by oxidative damage, at the molecular level, selenium has neuroprotective properties in the brain [105–108].

Several selenoproteins are expressed in the brain. Among them, according to the literature, the antioxidant effect of selenoprotein P on neuronal survival has been proven [109], and the role of neuronal selenoprotein is in the development of interneuronal connections and the prevention of seizures and the process of neurodegeneration [110]. However, its role in postischemic neuronal death cannot yet be explained.

With TBI, a reactive glial response is possible in the form of the development of astrogliosis-reactive gliosis in the hippocampus, and in the form of cellular hypertrophy, hyperplasia, increased release of glial fibrillar acid proteins.

The last study by O. Leiter et al. [111] has demonstrated that selenium mediates the exercise-induced increase in adult hippocampal neurogenesis, increases hippocampal precursor proliferation and adult neurogenesis, and reverses cognitive decline in aging and hippocampal injury [111].

Naziroglu et al. [9] did experimental work on rats, having created a hypoxic model of brain damage (convulsive seizures provoked by the administration of pentylenetetrazole). Selenium was preemptively injected at a dose of 0.3 mg intraperitoneally, then the activity of Ca++ -ATP-aza, the level of oxidative stress were measured for 7 days, and EEG was recorded in animals with affected brains [9]. The authors' conclusion: selenium caused protective effects on pentylenetetrazole-induced brain damage due to reduced production of free radicals, regulation of Ca++ – dependent processes, and maintenance of the antioxidant system.

The literature also mentions information that selenium deficiency in chickens caused a decrease in the activity of glutathione peroxidase, the level of expression of the mRNA glutathione peroxidase gene, the development of oxidative stress of brain tissue, hypothyroidism, alterations in ion profiles in chicken muscles, imbalance in Ca ++ homeostasis, and then morphological damage to nervous tissue [112].

In an experimental model of TBI in mice, analysis of key regulators of apoptosis during H2O2-induced apoptosis in cells showed that selenium blocks the activation of certain protein kinases (JNK)/38, triggering apoptosis in neuronal cells [6].

In vivo experiments have shown that selenite powerfully inhibits H2O2-induced apoptosis in neurons during TBI. Thus, selenium has protective properties and blocks apoptotic cell death, and participates in maintaining the functional activity of neurons and in the inhibition of astrogliosis [6].

Ozbal and colleagues [8] evaluated the levels of synthesis of tumor necrosis factor TNF-α and IL-1β, nerve tissue growth factor (NGF) in a cerebral ischemia/reperfusion model in rats [8]. In this study, they studied the effect of selenium on the prefrontal cortex and the degree of damage to the hippocampus in rats subjected to cerebral ischemia-reperfusion injury. Selenium was administered intraperitoneally to animals at a dose of 0.625 mg/kg/day after the onset of ischemic injury. Conclusion of the authors of the study: selenium treatment after ischemia significantly reduces the induced ischemia and subsequent reperfusion neuronal death in the prefrontal cortex and hippocampal CA 1 region in rats.

Selenium treatment reduces the levels of markers of systemic inflammatory response and tissue damage (TNF-α and IL-1β) and leads to an increase in the values of nerve tissue growth factor (NGF). B. Yang et al. [113] study was to explore the molecular mechanisms underlying the protective effects of selenium on the blood-brain barrier (BBB) following ischemia/reperfusion injury in hyperglycemic rats [113]. Treatment with selenium and the autophagy inhibitor 3-methyladenine significantly reduced cerebral infarct volume, brain water content, and Evans blue leakage, while increasing the expression of tight junction (TJ) proteins and decreasing that of autophagy-related proteins. It was revealed that selenium increased TJ protein levels, reduced BBB permeability, decreased autophagy levels, and enhanced the expression of phosphorylated (p)-AKT/AKT and p-mTOR/mTOR proteins [113].

In a study on mice, it was demonstrated that melatonin and selenium may serve as potential therapeutic targets against docetaxel-induced toxicity in the hippocampus and the brain (docetaxel is widely used to treat several types of glioblastoma) [114].

Summarizing the above, it can be argued that the results of experimental studies allow us to make the assumption that the introduction of selenium prevents the development of secondary pathological processes in the brain during its traumatic injury.

Clinicians, based on the data of experimental works performed on animals, can propose new goals of drug therapy for the treatment of TBI from the bench to the bedside.

5. Replacement selenium therapy in severe traumatic brain injury

Positive clinical responses obtained during therapy with N-acetylcysteine and selenium in neurodegenerative diseases have provided substantial evidence for the important role of reactive oxygen species in pathological processes of TBI [6]. It is proved that the level of oxidative stress in severe TBI determines the severity of the processes of necrobiosis and neuronal death [5].

Works concerning selenium metabolism in patients with severe trauma, including traumatic brain injury, are isolated [11, 23, 115–120].

In one study, a double-blinded controlled trial was carried out on 113 patients who were hospitalized following traumatic brain injury (TBI) with Glasgow Coma Scale score of 4–12 that were randomly assigned to receive selenium within 8 h after injury plus standard treatment group or routine standard treatment alone as the control. There was no difference in the length of ICU and hospital stay, the trend of the change in FOUR and SOFA scores within 15 days of first interventions, and the mean APACHE III score on the 1st and 15th days between the two groups. Mortality was 15.8% in the selenium group and 19.6% in the control group with no between-group difference. This human trial study could not demonstrate the beneficial effects of intravenous infusion of selenium in the improvement of outcomes in patients with acute TBI [120].

Several studies examine the effect of intravenous selenium (Selenase ®) treatment in patients with severe TBI on functional outcome and survival. Intravenous Selenase ® treatment demonstrates a significant improvement in functional neurologic outcomes [115]. H. S. Nutsalova in her study showed that selenium replacement therapy with Selenase ® at a dose of 1000 mcg/day for 12 days of the acute period of TBI significantly reduces the plasma level MDA (malonic aldehyde) in patients with severe TBI starting from day 7, reaching maximum intragroup and intergroup differences by day 12 ($p < 0,01$) [119]. Substitution selenium therapy does not affect the recovery time of consciousness in patients with severe TBI in the acute period of trauma. Replacement selenium therapy in patients with isolated and combined severe TBI provides the restoration of plasma levels of selenium and the sanogenetic orientation of free radical oxidation processes in the acute period of trauma. The known method of intravenous selenium use leads to a reduction in the duration of ventilation and a decrease in 28-day mortality in patients with severe TBI [116–119].

6. Nontraumatic acute cerebral damage

Hirato J et al. [121] demonstrated in their observation that the brain lesions of the megacystis microcolon intestinal hypoperistalsis syndrome (MMIHS) patients mainly resulted from oxidative damage of the brain related to the low levels of glutathione peroxidase and other selenoproteins due to selenium deficiency [121]. The authors showed that long-term total parenteral nutrition is possibly due to selenium deficiency. Both patients described in the article died of sepsis. In both cases, severe neuronal loss and gliosis were present in the medial convolutions of the occipital lobe, including the visual cortex [121].

Perinatal hypoxic-ischemic encephalopathy (HIE) is an important cause of brain injury in the newborn and can result in devastating consequences. The principle mechanism underlying neurological damage in HIE, resulting from hypoxemia and/or ischemia is deprivation of glucose and oxygen supply which energy failure. A consequent reperfusion injury often deteriorates the brain metabolism by increasing oxidative stress damage. Selenium is a constituent of the antioxidant enzyme glutathione peroxidase and is vital to antioxidant defense.

Neonates with HIE had lower serum selenium level than normal healthy neonates, which is not dependent on the maternal serum selenium levels and was negatively correlated with the severity of HIE [122].

Neonatal mortality continues to be a significant problem in the Indian setting, especially in very low birthweight (VLBW) neonates. India is a selenium-deficient country. Blood selenium concentrations in newborns are lower than those of their mothers and lower still in preterm infants.

Preterm VLBW neonates are selenium deficient at birth. Selenium supplementation at 10 µg/day resulted in getting the selenium levels into the acceptable normal level and reduced the incidence of the first episode of late-onset sepsis in these neonates [123].

7. Sepsis-associated encephalopathy and selenium status: perspectives of replacement therapy

Septis-associated encephalopathy is an early manifestation of systemic infection when the focus of infection is outside the central nervous system (CNS), but the systemic inflammatory response causes organ dysfunction, including the brain. Researchers identify a number of factors and mechanisms that play a key role in the development of septis-associated encephalopathy: the effect of inflammatory mediators on the brain, inadequate cerebral perfusion pressure, impaired permeability of the blood-brain barrier (BBB), disorders of the cerebral microcirculation, cerebral ischemia, metabolic disorders, changes in amino acid levels, imbalance of mediators, liver failure, and multiple organ failure [124, 125]. BBB dysfunction largely explains the pathophysiology of SAE, since the central nervous system becomes highly sensitive to neurotoxic factors, such as free radicals, inflammatory mediators, intravascular proteins, plasma, and circulating leukocytes. Due to the barrier deficiency, brain edema is formed and microvascular perfusion decreases, which leads to the loss of neurons during SAE [125].

Microglial cells are the primary inducers of immune responses in the brain. Recent experimental studies have shown that microglial cells migrate to brain vessels during systemic inflammation and that their activation represents one of the earliest changes observed in SAE [126, 127].

Designed to protect against sepsis, microglia activation generates cytotoxic substances that release reactive oxygen species (ROS), nitric oxide (NO), and glutamate SAE [127]. Persistent microglial activation and excessive release of inflammatory mediators and free radicals trigger a vicious cycle of a circle leading to the aberrant function of neurons and cell death, contributing to the progression of SAE [128]. Data from some experimental studies indicate that glial activation plays a key role in the development of SAE and BBB dysfunction along with a deficiency of brain neurotrophic factors [128, 129].

The pathophysiology of SAE is a multifactorial process that involves a violation of the mechanism of cell death. Ferroptosis is a new form of programmed cell death characterized by the accumulation of iron and lipid peroxidation, which leads to an inflammatory cascade and the release of glutamate. Scientists have suggested that ferroptosis is involved in glutamate-mediated excitotoxic damage to neurons during an uncontrolled inflammatory process in SAE [130].

Assessment of neurological status and neurocognitive deficit and their dynamics are the criteria for the effectiveness of treatment of neurocognitive disorders in patients with sepsis-associated encephalopathy, along with clinical and laboratory indicators and scales for assessing multiple organ failure (SAPS II, SOFA) [131]. Inflammatory cytokines and oxidative stress released during sepsis are high in septic patients, and their concentrations have some association with the severity and evolution of organ dysfunctions [132]. Decreased plasma selenium levels are found to be associated with excess mortality [133]. Plasma selenium concentrations in all patients with sepsis and septic shock are determined to be low (from 0.20 to 0.72 mcmol/l) [134].

Based on the understanding of the main mechanisms of selenium action—suppression of hyperactivation of NF-kB; reduction of activation of the complement system; immunomodulation and anti-inflammatory effect; maintenance of utilization of peroxides; suppression endothelial adhesion (decreased expression of E-selectin, P-selectin); protection of the endothelium from oxygen radicals (using selenoprotein P, which prevents the formation of peroxynitrite from O2 and NO) [129, 135], one can safely assume the expediency of using selenium-containing drugs in complex therapy of SAE to prevent the development of neurocognitive deficiency due to the mechanisms of neuroinflammation in the future [124].

8. Conclusion

The role of selenium in metabolic processes and the importance of correcting its level in various diseases and critical conditions are widely covered in modern literature [40, 43–48, 50, 70, 135]. Selenium deficiency, which occurs during the development of oxidative stress due to severe TBI, sepsis, and other critical conditions, significantly affects the work of antioxidant systems, reduces the protective mechanisms of the patient's body and requires correction.

The results of studies of selenium homeostasis in *in vivo* experiments and in the human body in normal and various pathological conditions obtained over the past 20 years indicate the direct participation of this most important nutrient in the body's defense mechanisms in severe trauma, burns, sepsis, TBI, acute cerebral injury of nontraumatic etiology, etc. Correction of the selenium status of patients in critical condition is especially relevant, since selenium deficiency blocks adequate antioxidant protection, a full-fledged immune response to infection, and relief of excessive systemic inflammatory response and the integrative complex response of the brain to any damaging effect that poses a threat to the survival of a mammal.

We are confident that the importance of selenium deficiency correction in the form of selenium replacement therapy is reflected in the treatment protocols of patients in critical conditions, including acute cerebral injury.

Conflict of interest

The authors declare no conflict of interest.

Abbreviation

ALT	alanine aminotransferase
APACHE III	Acute Physiology and Chronic Health Evaluation III score
AST	aspartate aminotransferase
ATP	adenosine triphosphate
BBB	blood-brain barrier
DiO2	Iodothyronine deiodinase 2
FOUR	Full Outline of UnResponsiveness Score
HIE	hypoxic-ischemic encephalopathy
H_2S	hydrogen sulfide
ICU	intensive care unite
IL-1β	interleukin 1β
GCR	glucocorticoid receptors
GPX1-6	glutathione peroxidases 1-6
LDH	lactate dehydrogenase
mRNA	mitochondrial ribonucleic acid
Na2SeO3	sodium selenite
ROS	reactive oxygen species
SAE	sepsis-associated encephalopathy
SOFA	Sequential Organ Failure Assessment Score
TBI	traumatic brain injury
TNF-α	tumor necrosis factor-alpha

Author details

Irina Alexandrovna Savvina[1*], Hasaybat Salimbekovna Nucalova[2], Anna Olegovna Petrova[1], Kristina M. Bykova[1] and Irina Varlamovna Tkebuchava[3]

1 Almazov National Medical Research Centre of Ministry of Health Care of Russia, Saint Petersburg, Russian Federation

2 Republican Clinical Hospital, Makhachkala Town, The Republic of Dagestan, Russian Federation

3 Samara City N. I. Pirogov Clinical Hospital, Samara Town, Russian Federation

*Address all correspondence to: irinasavvina@mail.ru

References

[1] Chen J, Berry MJ. Selenium and selenoproteins in the brain and brain diseases. Journal of Neurochemistry. 2007;**86**:1-12

[2] Rayman MP. The importance of selenium to human health. Lancet. 2000;**356**:233-241

[3] Rao L, Puschner B, Prolia TA. Gene expression profiling of low selenium status in the mouse intestine. The Journal of Nutrition. 2001;**131**:3175-3181

[4] Vladimirov YA. Free radicals in biological systems. Yu. A. Vladimirov Biology. 2000;**6**(12):13-19

[5] Zaichik AS, Churilov LP. General Pathophysiology. St. Petersburg: Albi; 2001. p. 467

[6] Yeo JE, Kyung S, Kang SK. Selenium effectively inhibits ROS-mediated apoptotic neural precursor cell death in vitro and in vivo in traumatic brain injury. Journal of Biochimica et Biophysica Acta. 2007;**1772**(11-12):1199-1210

[7] Kiymaz N, Ekin S, Yilmaz N. Plasma vitamin E and selenium levels in rats with head trauma. Surgical Neurology. 2007;**68**(1):67-70

[8] Ozbal S, Erbil G, Kocdor H, Tugyan K, Pekcetin C, Ozogul C. The effects of selenium against cerebral ischemia-reperfusion injury in rats. Neuroscience Letters. 2008;**438**(3):265-269

[9] Naziroglu M, Kutluhan S, Yilmaz M. Selenium and topizamate modulates brain microsomal oxidative stress values, Ca 2+-ATPase activity, and EEG records in pentylentetrazol-induced seizures in rats. The Journal of Membrane Biology. 2012;**225**(1):39-49

[10] Dalla Puppa L, Savaskan NE, Bräuer AU, Behne D, Kyriakopoulos A. The role of selenite on microglial migration. Annals of the New York Academy of Sciences. 2007;**1096**:179-183

[11] Kaziakhmedov VA. Selenium metabolism in severe TBI pediatric patients. Abstract of the dissertation of a candidate of medical sciences. Saint-Petersburg. 2007:19

[12] Savaskan NE, Bräuer AU, Kühbacher M, Eyüpoglu IY, Kyriakopoulos A, Ninnemann O, et al. Selenium deficiency increases susceptibility to glutamate-induced excitotoxicity. The FASEB Journal. 2003;**17**:112-114

[13] Hill KE, Zhou J, Austin LM, Motley AK, Ham AJ, Olson GE, et al. The selenium-rich C-terminal domain of mouse selenoprotein P is necessary for the supply of selenium to brain and testis but not for the maintenance of whole body selenium. The Journal of Biological Chemistry. 2007;**282**:10972-10980

[14] Scharpf M, Schweizer U, Arzberger T, Roggendorf W, Schomburg L, Kohrle J. Neuronal and ependymal expression of selenoprotein P in the human brain. Journal of Neural Transmission. 2007;**114**:877-884

[15] Schomburg L, Schweizer U, Holtmann B, Flohe L, Sendtner M, Kohrle J. Gene disruption discloses role of selenoprotein P in selenium delivery to target tissues. The Biochemical Journal. 2003;**370**:397-402

[16] Burk RF, Hill KE. Selenoprotein P-expression, functions, and roles in mammals. Biochimica et Biophysica Acta. 2009;**1790**(11):1441-1447

[17] Torres DJ, Alfulaij N, Berry MJ. Stress and the brain: An emerging role for selenium. Frontiers in Neuroscience. 2021;**15**:666601. DOI: 10.3389/fnins.2021.666601

[18] Solovyev ND. Importance of selenium and selenoprotein for brain function: From antioxidant protection to neuronal signalling. Journal of Inorganic Biochemistry. 2015;**153**:1-12

[19] Nakashidze I, Chikovani T, Sanikidze T, Bakhutashvili V. Manifestations of oxidative stress and its correction in traumatic shock. Anesthesiology and resuscitation. 2003;**5**:22-24

[20] Alonso de Vega JM, Diaz J, Serrano E, Carbonell LF. Oxidative stress in critically ill patients with systemic inflammatory response syndrome. Critical Care Medicine. 2002;**30**:1782-1786

[21] Motoyama T, Okamoto K, Kukita I, Hamaguchi M, Kinoshita Y, Ogawa H. Possible role of increased oxidant stress in multiple organ failure after systemic inflammatory response syndrome. Critical Care Medicine. 2003;**31**:1048-1052

[22] Avenell A, Noble DW, Barr J, Engelhardt T. Selenium supplementation for critically Ill adults. Anesthesia and Analgesia. 2005;**100**(5):1536-1536

[23] Aleksandrovich YS, Kaziakhmedov VA, Arutsova IY, Pshenisnov KV, Korchagin IV. Dynamics of changes in selenium concentration in plasma, urine and erythrocytes in children with severe traumatic brain injury. Anesthesiology and Resuscitation. 2008;**1**:23-26

[24] Huang Z, Rose AH, Hoffmann PR. The role of selenium in inflammation and immunity: From molecular mechanisms to therapeutic opportunities. Antioxidants & Redox Signaling. 2012;**16**(7):705-743

[25] Sakr Y, Reinhart K, Bloos F, Marx G, Russwurm S, Bauer M, et al. Time course and relationship between plasma selenium concentrations, systemic inflammatory response, sepsis, and multiorgan failure. British Journal of Anaesthesia. 2007;**98**(6):775-784

[26] Andrews P. Selenium and glutamine supplements: Where are we heading? A critical care perspective. Current Opinion in Clinical Nutrition and Metabolic Care. 2010;**13**:192-197

[27] Andrews PJD, Avenell A, Noble DW, Campbell MK, Croal BL, Simpson WG, et al. Randomised trial of glutamine, selenium, or both, to supplement parenteral nutrition for critically ill patients. BMJ. 2011;**342**:d1542. DOI: 10.1136/bmj.d1542

[28] Berger MM. Antioxydant micronutrients in major trauma and burns: Evidence and practice. Nutrition in Clinical Practice. 2006;**21**(5):438-449

[29] Tutelyan VA, Knyazhev VA, Khotimchenko SA, Golubkina NA, Kushlinsky NE, Sokolov YA. Selenium in the Human Body. Moscow: RAMS Publishing House; 2002. p. 219

[30] Maehira F, Miyagi I, Eguchi Y. Selenium regulates transcription factor NF-kB activation during the acute phase reaction. Clinica Chimica Acta. 2003;**334**:163-171

[31] Geoghegan M, McAnley D, Eaton S, Powell-Tuck J. Selenium in critical illness: Current opinion. Critical Care. 2006;**12**(2):136-141

[32] Angstwurm MW, Schottdorf J, Schopohl J, Gaertner R. Selenium replacement in patients with severe

systemic inflammatory response syndrome improves clinical outcome. Critical Care Medicine. 1999;**27**:1807-1813

[33] Angstwurm MW. Practicalities of selenium supplementation in critically ill patients. Current Opinion in Clinical Nutrition and Metabolic Care. 2006;**9**(3):233-238

[34] Angstwurm MW, Engelmann L, Zimmermann T, Lehmann C, Spes CH, Abel P, et al. Selenium in intensive care (SIC): Results of a prospective randomized, placebo-controlled, multiple-center study in patients with severe systemic inflammatory response syndrome, sepsis, and septic shock. Critical Care Medicine. 2007;**35**:118-126

[35] Manoli I, Alesci S, Blackman MR, Su YA, Rennert OM, Chrousos GP. Mitochondria as key components of the stress response. Trends in Endocrinology and Metabolism. 2007;**18**:190-198

[36] Mehta SL, Kumari S, Mendelev N, Li PA. Selenium preserves mitochondrial function, stimulates mitochondrial biogenesis, and reduces infarct volume after focal cerebral ischemia. BMC Neuroscience. 2012;**13**:79

[37] Mendelev N, Mehta SL, Witherspoon S, He Q, Sexton JZ, Li PA. Upregulation of human selenoprotein H in murine hippocampal neuronal cells promotes mitochondrial biogenesis and functional performance. Mitochondrion. 2011;**11**:76-82

[38] Forceville X. Effects of high doses of selenium, as sodium selenite, in septic shock patients a placebocontrolled, randomized, double-blind, multi-center phase II study-selenium and sepsis. Journal of Trace Elements in Medicine and Biology. 2007;**21**(Suppl 1):62-65

[39] Forceville X, Laviolle B, Annane D, Vitoux D, Bleichner G, Korach JM, et al. Effects of high doses of selenium, as sodium selenite, in septic shock: A placebo controlled, randomized, double-blind, phase II study. Critical Care. 2007;**11**:R73

[40] Vincent JL, Forceville X. Critically elucidating the role of selenium. Current Opinion in Anaesthesiology. 2008;**21**:148-154

[41] Gelfand BR, Yakovleva II, Popov TV, Glushko AV. Experience of using the drug Selenase in the intensive care complex of patients with destructive pancreatitis. Infections in Surgery. 2008;**1**:54-56

[42] Wang Z, Forceville X, Van Antwerp P, Piagnerelli M, Ahishakiye D, Macours P, et al. A large bolus, but not a continuous infusion, of sodium selenite improves outcome in peritonitis. Shock. 2009;**32**:140-146

[43] Manzanares W, Biestro A, Torre MH, Galusso F, Facchin G, Hardy G. High-dose selenium reduces ventilator-associated pneumonia and illness severity in critically ill patients with systemic inflammation. Intensive Care Medicine. 2011;**37**:1120-1127

[44] Manzanares W, Biestro A, Torre MH, Galusso F, Forre MN, Manay N, et al. Serum selenium and glutathion peroxidase-3 activity: biomarkers of systemic inflammation in the critically ill. Intensive Care Medicine. 2009;**35**:882-889

[45] Manzanares W, Hardy G. Selenium supplementation in the critically ill. Phisiology and pharmacokinetics. Current Opinion in Clinical Nutrition and Metabolic Care. 2009;**12**:273-280

[46] Manzanares W., Biestro A., Galusso F. et al. Serum selenium and glutationeperoxidase-3 activity biomarkers of systemic inflammation in the critically ill. Intensive Care Medicine. 2009; 35. p. 882-889

[47] Manzanares W, Biestro A, Torre MH, Galusso F, Facchın G, Hardy G. Clinical effects of high dose selenious acid in critically ill patients with systemic inflammation. Journal of Parenteral and Enteral Nutrition. 2009;**33**:186

[48] Manzanares W, Biestro A, Torre MH, Galusso F, Forre MN, Manay N, et al. High-dose selenium for critically ill patients with systemic inflammation: A pilot study. Nutrition. 2010;**26**:634-640

[49] Tinggi U. Selenium: Its role as antioxidant in human health. Environmental Health and Preventive Medicine. 2008;**13**:102-108

[50] Matthias WA, Angstwurm MW, Engelmann L, Zimmermann T, et al. Selenium in Intensive Care (SIC) study: Results of a prospective randomized, placebo-controlled, multiple-center study in patients with severe systemic inflammatory response syndrome, sepsis, and septic shock. Critical Care Medicine. 2007;**35**(1):1-8

[51] Koszta G, Kacska Z, Szatmari K, Szerafin T, Fulesdi B. Lower whole blood selenium level is associated with higher operative risk and mortality following cardiac surgery. Journal of Anesthesia. 2012;**26**:812-821

[52] Forceville X, Van Antwerpen P, Annane D, Vincent JL. Selenocompounds and Sepsis-Redox bypass hypothesis: Part B-Selenocompounds in the management of early Sepsis. Antioxidants & Redox Signaling. 2022;**37**(13-15):998-1029

[53] Gao Y, Wang Z, Li Y, Yang J, Liao Z, Liu J, et al. A rational design of copper–selenium nanoclusters that cures sepsis by consuming endogenous H2S to trigger photothermal therapy and ROS burst. Biomaterials Science. 2022;**10**(12):3137-3157

[54] Lehmann C, Egerer K, Weber M, Krausch D, Wauer H, Newie T, et al. Effect of selenium administration on various laboratory parameters of patients at risk for sepsis syndrome. Medizinische Klinik. 2007;**15**(Suppl. 3):14-16

[55] Shenkin A. Selenium in intravenous nutrition. Gastroenterology. 2009;**137**:S61-S69

[56] Singer P, Berger MM, Van den Berghe G, Biolo G, Calder P, Forbes A, et al. ESPEN Guidelines on parenteral nutrition: Intensive care. Clinical Nutrition. 2009;**28**:387-400

[57] Bozzetti F, Forbes A. The ESPEN clinical practice guidelines on parenteral nutrition: present status and perspectives for future research. Clinical Nutrition. 2009;**28**:359-364

[58] Rupinder D, Naomi C, Margot L, Daren KH. The Canadian critical care nutrition guidelines in 2013: An update on current recommendations and implementation strategies. Nutrition in Clinical Practice. 2014;**29**(1):29-43. DOI: 10.1177/0884533613510948

[59] von Gagern G, Zimmermann T, Albrecht S, et al. Significance of selenium in regulation of inflammatory response b transcription factors in polytrauma patients. A clinical study. Medizinische Klinik. 1999;**15**(Suppl 3):62-65

[60] Nazarov RV, Tsentsiper LM, Kondratieva EA, Kondratiev SA, Semenov EL, Dryagina NV. The place and role of selenium in the correction of systemic inflammatory response of patients in prolonged unconsciousness. Efferent Therapy. 2011;**17**(3):100-101

[61] Chelkeba L, Ahmadi A, Abdollahi M, et al. The effect of parenteral selenium on outcomes of mechanically ventilated patients following sepsis: A prospective

randomized clinical trial. Annals of Intensive Care. 2015;**5**:29. DOI: 10.1186/s13613-015-0071-y

[62] Landesberg G, Levin PD, Gilon D, et al. Disfunção miocárdica em sepse grave e choque séptico—sem correlação com citocinas inflamatórias em ambiente clínico da vida real. Baú. 2015;**148**(1):93-102

[63] Kieliszek M, Lipinski B. Selenium supplementation in the prevention of coronavirus infections (COVID-19). Medical Hypotheses. 2020;**143**:109878. DOI: 10.1016/j.mehy.2020.109878

[64] Mahmoodpoor A, Hamishehkar H, Sanaie S, et al. Antioxidant reserve of the lungs and ventilator-associated pneumonia: A clinical trial of high dose selenium in critically ill patients. Journal of Critical Care. 2018;**44**:357-362. DOI: 10.1016/j.jcrc.2017.12.016

[65] Huang JF, Hsu CP, Ouyang CH, Cheng CT, Wang CC, Liao CH, et al. The impact of selenium supplementation on trauma patients-systematic review and Meta-analysis. Nutrients. 2022;**14**(2):342

[66] Kong L, Wu Q, Liu B. The impact of selenium administration on severe sepsis or septic shock: A meta-analysis of randomized controlled trials. African Health Sciences. 2021;**21**(1):277-285

[67] Manzanares W, Dhaliwal R, Jiang X, Murch L, Heyland DK. Antioxidant micronutrients in the critically ill: A systematic review and meta-analysis. Critical Care. 2012;**16**(2):R66

[68] Valenta J, Brodska H, Drabek T, Hendl J, Kazda A. High-dose selenium substitution in sepsis: A prospective randomized clinical trial. Intensive Care Medicine. 2011;**37**(5):808-815

[69] Zhang X, Liu RP, Cheng WH, Zhu JH. Prioritized brain selenium retention and selenoprotein expression: Nutritional insights into Parkinson's disease. Mechanisms of Ageing and Development. 2019;**180**:89-96

[70] Solovyev N, Drobyshev E, Bjørklund G, Dubrovskii Y, Lysiuk R, Rayman MP. Selenium, selenoprotein P, and Alzheimer's disease: Is there a link? Free Radical Biology & Medicine. 2018;**127**:124-133

[71] de Toledo JHDS, de TF Fraga Silva C, Borim PA, de Oliveira LRC, da Oliveira ES, Périco LL. Organic selenium reaches the central nervous system and downmodulates local inflammation: A complementary therapy for multiple sclerosis? Frontiers in Immunology. 2020;**11**:571844

[72] Alim I, Caulfield JT, Chen Y, Swarup V, Geschwind DH, Ivanova E, et al. Selenium drives a transcriptional adaptive program to block ferroptosis and treat stroke. Cell. 2019;**177**(5):1262-1279.e25

[73] Meseguer I, Molina JA, Jimenez-Jimenez FJ, et al. Cerebrospinal fluid levels of selenium in patients with Alzheimer's disease. Journal of Neural Transmission. 1999;**106**:309-315

[74] Gao S, Jin Y, Hall KS, Liang C, Unverzagt FW, Ji R, et al. Selenium level and cognitive function in rural elderly Chinese. American Journal of Epidemiology. 2007;**165**:955-965

[75] Thiel R, Fowkes SW. Can cognitive deterioration associated with Down syndrome be reduced? Medical Hypotheses. 2005;**64**(3):524-532

[76] Zafar KS, Siddiqui A, Sayeed I, Ahmad M, Salim S, Islam F. Dose-dependent protective effect of selenium

in rat model of Parkinson's disease: Neurobehavioral and neurochemical evidences. Journal of Neurochemistry. 2003;**84**:438-446

[77] Aguilar MV, Jimenez-Jimenez FJ, Molina JA, et al. Cerebrospinal fluid selenium and chromium levels in patients with Parkinson's disease. Journal of Neural Transmission. 1998;**105**:1245-1251

[78] Kim H, Jhoo W, Shin E, Bing G. Selenium deficiency potentiates methamphetamine-induced nigral neuronal loss; comparison with MPTP model. Brain Research. 2000;**862**: 247-252

[79] Kim HC, Jhoo WK, Choi DY, Im DH, Shin EJ, Suh JH, et al. Protection of methamphetamine nigrostriatal toxicity by dietary selenium. Brain Research. 1999;**851**:76-86

[80] Steinbrenner H, Alili L, Bilgic E, Sies H, Brenneisen P. Involvement of selenoprotein P in protection of human astrocytes from oxidative damage. Free Radical Biology & Medicine. 2006;**40**:1513-1523

[81] Ashrafi MR, Shams S, Nouri M, Mohseni M, Shabanian R, Yekaninejad MS, et al. A probable causative factor for an old problem: Selenium and glutathione peroxidase appear to play important roles in epilepsy pathogenesis. Epilepsia. 2007;**13**:1256-1268

[82] Nazıroğlu M, Yürekli VA. Effects of antiepileptic drugs on antioxidant and oxidant molecular pathways: Focus on trace elements. Cellular and Molecular Neurobiology. 2013;**33**(5):589-599

[83] Hardy G, Hardy I. Selenium: The Se-XY nutraceutical. Nutrition. 2004;**20**:590-593

[84] Zimmermann C, Winnefeld K, Streck S, Roskos MM, Haberl RL. Antioxidant status in acute stroke patients and patients at stroke risk. European Neurology. 2004;**51**(3): 157-161. DOI: 10.1159/000077662

[85] Ansari MA, Ahmad AS, Ahmad M, Salim S, Yousuf S, Ishrat T, et al. Selenium protects cerebral ischemia in rat brain mitochondria. Biological Trace Element Research. 2004;**101**:73-86

[86] Arakawa M, Yoshihiso I. N-acetylcysteine and neurodegenerative diseases: Basic and clinical pharmacology. Cerebellum. 2007;**6**(4):308-314

[87] Yousuf S, Atif F, Ahmad M, Hoda MN, Khan MB, Ishrat T, et al. Selenium plays a modulatory role against cerebral ischemia-induced neuronal damage in rat hippocampus. Brain Research. 2007;**1147**:218-225

[88] Wray JR, Davies A, Sefton C, Allen TJ, Adamson A, Chapman P, et al. Global transcriptomic analysis of the arcuate nucleus following chronic glucocorticoid treatment. Molecular Metabolism. 2019;**26**:5-17

[89] Perez A, Jansen-Chaparro S, Saigi I, Bernal-Lopez MR, Miñambres I, Gomez-Huelgas R. Glucocorticoid-induced hyperglycemia. Journal of Diabetes. 2014;**6**(1):9-20

[90] Saito I, Asano T, Sano K, Takakura K, Abe H, Yoshimoto T, et al. Neuroprotective effect of an antioxidant, ebselen, in patients with delayed neurological deficits after aneurysmal subarachnoid hemorrhage. Neurosurgery. 1998;**42**:269-277

[91] Yamaguchi T, Sano K, Takakura K, Saito I, Shinohara Y, Asano T, et al. Ebselen in acute ischemic stroke: A placebo-controlled, double-blind clinical

trial. Ebselen Study Group. Stroke. 1998;**29**:12-17

[92] Kühbacher M, Bartel J, Hoppe B, Alber D, Bukalis G, Bräuer A, et al. The brain selenoproteome: Priorities in the hierarchy and different levels of selenium homeostasis in the brain of selenium-deficient rats. Journal of Neurochemistry. 2009;**110**:133-142

[93] Xu L, Zhang S, Chen W, Yan L, Chen Y, Wen H, et al. Trace elements differences in the depression sensitive and resilient rat models. Biochemical and Biophysical Research Communications. 2020;**529**(2):204-209

[94] Pasco JA, Jacka FN, Williams LJ, Evans-Cleverdon M, Brennan SL, Kotowicz MA, et al. Selenium and major depression: A nested case-control study. Complementary Therapies in Medicine. 2012;**20**(3):119-123

[95] Burk RF, Hill KE. Selenoprotein P: An extracellular protein with unique physical characteristics and a role in selenium homeostasis. Annual Review of Nutrition. 2005;**25**:215-235

[96] Burk RF, Hill KE, Olson GE, Weeber EJ, Motley AK, Winfrey VP, et al. Deletion of apolipoprotein E receptor-2 in mice lowers brain selenium and causes severe neurological dysfunction and death when a low-selenium diet is fed. The Journal of Neuroscience. 2007;**27**:6207-6211

[97] Hoffmann PR, Höge SC, An LP, Hoffmann FW, Hashimoto AC, Berry MJ. The selenoproteome exhibits widely varying, tissue-specific dependence on selenoprotein P for selenium supply. Nucleic Acids Research. 2007;**35**(12):3963-3973. DOI: 10.1093/nar/gkm355

[98] Erbil G, Ozbal S, Sonmez U, Pekketin C, Tugyan K, Bayriyanik A, et al. Neuroprotective effects of selenium and ginkgo biloba extract (Egb761) against ischemia and reperfusion injury in rat brain. Neurosciences. 2008;**13**(3):233-238

[99] Casaril AM, Domingues M, Bampi SR, de Andrade LD, Padilha NB, Lenardão EJ, et al. The selenium-containing compound 3-((4-chlorophenyl)selanyl)-1-methyl-1H-indole reverses depressive-like behavior induced by acute restraint stress in mice: Modulation of oxido-nitrosative stress and inflammatory pathway. Psychopharmacology (Berl). 2019;**236**(10):2867-2880

[100] Casaril AM, Ignasiak MT, Chuang CY, Vieira B, Padilha NB, Carroll L, et al. Selenium-containing indolyl compounds: Kinetics of reaction with inflammation-associated oxidants and protective effect against oxidation of extracellular matrix proteins. Free Radical Biology & Medicine. 2017;**113**:395-405

[101] Casaril AM, Domingues M, Fronza M, Vieira B, Begnini K, Lenardão EJ, et al. Antidepressant-like effect of a new selenium-containing compound is accompanied by a reduction of neuroinflammation and oxidative stress in lipopolysaccharide-challenged mice. Journal of Psychopharmacology (Oxford, England). 2017;**31**(9):1263-1273

[102] Yeo JE, Kim JH, Kang SK. Selenium attenuates ROS-mediated apoptotic cell death of injured spinal cord through prevention of mitochondria dysfunction; in vitro and in vivo study. Cellular Physiology and Biochemistry. 2008;**21**:225-238

[103] Yin W, Signore AP, Iwai M, Cao G, Gao Y, Chen J. Rapidly increased neuronal mitochondrial biogenesis after hypoxic-ischemic brain injury. Stroke. 2008;**39**:3057-3063

[104] Zhou YJ, Zhang SP, Liu CW, Cai YQ. The protection of selenium on ROS mediated-apoptosis by mitochondria dysfunction in cadmium-induced LLC-PK(1) cells. Toxicology In Vitro. 2009;**23**:288-294

[105] Schweizer U, Brauer AU, Kohrle J, Nitsch R, Savaskan NE. Selenium and brain function: a poorly recognized liaison. Brain Research. Brain Research Reviews. 2004;**45**:164-178

[106] Schweizer U, Schomburg L. Selenium, selenoproteins and brain function. In: Hatfield DL, Berry MJ, Gladyshev VN, editors. Selenium its Molecular Biology and Role in Human Health. US: Springer; 2004. pp. 233-248

[107] Wang Q, Zhang QG, Wu DN, Yin XH, Zhang GY. Neuroprotection of selenite against ischemic brain injury through negatively regulating early activation of ASK1/JNK cascade via activation of PI3K/AKT pathway. Acta Pharmacologica Sinica. 2007;**28**:19-27

[108] Sarada SK, Himadri P, Ruma D, Sharma SK, Pauline T. Mrinalini: Selenium protects the hypoxia induced apoptosis in neuroblastoma cells through upregulation of Bcl-2. Brain Research. 2008;**1209**:29-39

[109] Peters MM, Hill KE, Burk RF, Weeber EJ. Altered hippocampus synaptic function in selenoprotein P deficient mice. Molecular Neurodegeneration. 2006;**1**:12

[110] Wirth EK, Conrad M, Winterer J, Wozny C, Carlson BA, Roth S, et al. Neuronal selenoprotein expression is required for interneuron development and prevents seizures and neurodegeneration. The FASEB Journal. 2010;**24**:844-852

[111] Leiter O, Zhuo Z, Rust R, Wasielewska JM, Gronnert L, Kowal S, et al. Selenium mediates exercise-induced adult neurogenesis and reverses learning deficits induced by hippocampal injury and aging. Cell Metabolism. 2022;**34**:408-423

[112] Yao H, Zhao X, Fan R, Sattar H, Zhao J, Zhao W, et al. Selenium deficiency-induced alterations in ion profiles in chicken muscle. PLoS One. 6 Sep 2017;**12**(9):e0184186. DOI: 10.1371/journal.pone.0184186

[113] Yang B, Li Y, Ma Y, Zhang X, Yang L, Shen X, et al. Selenium attenuates ischemia/reperfusion injury-induced damage to the blood-brain barrier in hyperglycemia through PI3K/AKT/mTOR pathway-mediated autophagy inhibition. International Journal of Molecular Medicine. 2021;**48**(3):1-13

[114] Ataizi ZS, Ertilav K, Nazıroğlu M. Mitochondrial oxidative stress-induced brain and hippocampus apoptosis decrease through modulation of caspase activity, Ca2+ influx and inflammatory cytokine molecular pathways in the docetaxel-treated mice by melatonin and selenium treatments. Metabolic Brain Disease. 2019;**34**(4):1077-1089

[115] Khalili H, Ahl R, Cao Y, Paydar S, Sjölin G, Niakan A, et al. Early selenium treatment for traumatic brain injury: Does it improve survival and functional outcome? Injury. 2017;**48**(9):1922-1926

[116] Savvina IA. Effective criteria of selenium therapy in patients with brain injury. I.A. Savvina, H.S. Nucalova. 6 th International Baltic Congress Anaesthesiology Intensive Care. Acta Medica Lituanica. 2012;**19**(3):391

[117] Savvina, I.A. Selenium therapy controls the oxidative stress level in patients with traumatic brain injury. I.A.

Savvina, H.S. Nucalova, M. D. Astaeva 32 nd Congress of the Scandinavian Society of Anaesthesiology and Intensive Care Medicine. Acta Anaesthesiologica Scandinavica. 2013;**57**(Suppl. 120):30

[118] Nutsalova HS, Mishina TP, Savvina IA. Dynamics of some clinical and laboratory indicators of systemic inflammatory response in patients with severe traumatic brain injury depending on the intake of selenium-containing drugs. Emergency Medical Help. 2014;**15**(2):60-64

[119] Nutsalova HS. Selenium replacement therapy in severe TBI patients. Abstract of the dissertation of a candidate of medical sciences. Makhachkala: Dzhamaludinov Publisher House; 2014. p. 24

[120] Moghaddam OM, Lahiji MN, Hassani V, Mozari S. Early administration of selenium in patients with acute traumatic brain injury: A randomized double-blinded controlled trial. Indian Journal of Critical Care Medicine: Peer-reviewed, Official Publication of Indian Society of Critical Care Medicine. 2017;**21**(2):75-79

[121] Hirato J, Nakazato Y, Koyama H, Yamada A, Suzuki N, Kuroiwa M, et al. Encephalopathy in megacystis-microcolon-intestinal hypoperistalsis syndrome patients on long-term total parenteral nutrition possibly due to selenium deficiency. Acta Neuropathol (Berl). 2003;**106**(3):234-242

[122] Abdel-Azeem M. El-Mazary, Reem A. Abdel-Aziz, Ramadan A. Mahmoud, Mostafa A. El-Said & Nashwa R. Mohammed correlations between maternal and neonatal serum selenium levels in full term neonates with hypoxic ischemic encephalopathy. Italian Journal of Pediatrics. 2015;**41**:83

[123] Aggarwal R, Gathwala G, Yadav S, Kumar P. Selenium supplementation for prevention of late-onset Sepsis in very low birth weight preterm neonates. Journal of Tropical Pediatrics. 2016;**62**(3):185-193

[124] Bykova KM, Savvina IA, Zabrodskaya YM, Bodareva NV. Pathophysiological aspects and complex diagnostics of sepsis-associated encephalopathy. Perspectives of etiopathogenetic therapy. Anesthesiology and Reanimatology. 2022;**4**:92-98

[125] Sonneville R, Verdonk F, Rauturier C, et al. Understanding brain dysfunction in sepsis. Annals of Intensive Care. 2013;**3**:15

[126] Haruwaka K, Ikegami A, Tachibana Y, et al. Dual microglia effects on blood brain barrier permeability induced by systemic inflammation. Nature Communications. 2019;**10**:5816. DOI: 10.1038/s41467-019-13812-z

[127] Erickson MA, Banks WA. Neuroimmune axes of the blood-brain barriers and blood-brain interfaces: Bases for physiological regulation, disease states, and pharmacological interventions. Pharmacological Reviews. 2018;**70**:278-314

[128] Wang H, Wang H, Song Y, et al. Overexpression of Foxc1 ameliorates sepsis associated encephalopathy by inhibiting microglial migration and neuroinflammation through the IκBα/NFκB pathway. Molecular Medicine Reports. 2022;**25**:107

[129] Solovyev N, Drobyshev E, Blume B, Michalke B. Selenium at the neural barriers: A review. Frontiers in Neuroscience. 2021, 2021;**15**:630016. DOI: 10.3389/fnins.2021.630016

[130] Xie Z, Xu M, Xie J, et al. Inhibition of ferroptosis attenuates glutamate

excitotoxicity and nuclear autophagy In a CLP septic mouse model. Shock Augusta Ga. 2022;**70**:278

[131] Savvina IA, Ryzhkova DV, Bykova KM, Lebedev KE, Petrova AO, Dryagina NV, et al. Diagnostics of central and autonomic nervous system dysfunction in patients with Sepsis-associated encephalopathy. In: Huang L, Zhang Y, Sun L, editors. Sepsis New Perspectives. London: IntechOpen; 2022. DOI: 10.5772/intechopen.108392. Available from: https://www.intechopen.com/online-first/diagnostics-of-central-and-autonomic-nervous-system-dysfunction-in-patients-with-sepsis-associated-e

[132] Eidt MV, Nunes FB, Pedrazza L, Caeran G, Pellegrin G, Melo DA, et al. Biochemical and inflammatory aspects in patients with severe sepsis and septic shock: The predictive role of IL-18 in mortality, Clinica chimica acta. International Journal of Clinical Chemistry. 2016;**453**:100-106

[133] Jang JY, Shim H, Lee SH, Lee JG. Serum selenium and zinc levels in critically ill surgical patients. Journal of Critical Care. 2014;**29**(2):317-318

[134] Forseville X, Vitox D, Gauzitetal R. Selenium, systemic immune response syndrome, sepsis, and outcome in critically ill patients. Critical Care Medicine. 1998;**26**(9):1536-1544

[135] Panee J, Liu W, Nakamura K, Berry MJ. The responses of HT22 cells to the blockade of mitochondrial complexes and potential protective effect of selenium supplementation. International Journal of Biological Sciences. 2007;**3**:335-341

Chapter 7

Increased Morbidity and Its Possible Link to Impaired Selenium Status

Shukurlu Yusif Hajibala and Huseynov Tokay Maharram

Abstract

This chapter summarizes the latest information on the main differences in the chemical properties of selenium proteins and their sulfur analogues, Se proteins and their functions, Se-accumulating proteins, the relationship between Se and hemoglobin, Selenium in gerontology, Selenium and iodine deficiency conditions, Se and immunity, Selenium as an antioxidant in nitrite poisoning. Also discussed are some of the results of the first studies on protein enrichment with selenium carried out in the seventies of the last century. This native protein was natural silk fibroin. Fibroin has since become an important tool for human health and healing. It was discovered that when selenium-containing inorganic compounds were added to mulberry silkworm feed, selenium atoms formed additional sulfur-like bonds in fibroin macromolecules. This resulted in additional branching of protein macromolecule. Selenium atoms in the fibroin structure have a sufficiently high electron affinity, act as small traps and capture migrating electrons. This leads to a reduction of free radicals, which are generated by external influences such as mechanical, thermal, electrical and radiation.

Keywords: selenium, hemoglobin, erythrocyte, *HbBcys 93*, nitric oxide, nitrite, COVID 19, viral diseases, fibroin, selenium enrichment

1. Introduction

One of the trace elements, the lack of which has a significant impact on human health, is selenium (*Se*). It is a part of many proteins and key antioxidant enzymes involved in many metabolic processes and has antioxidant and immunoregulatory properties. Its deficiency leads, first, to the weakening of the antioxidant defense system and immunity, which causes the development of several diseases. The content of selenium in the human body depends on the level of its dietary intake, which is closely related to the distribution of the element in the biosphere of the region of residence. At status of selenium in Azerbaijan, as well as in many other countries, is close to deficiency, and its decrease is connected with the worsening of the ecological situation. The problem of selenium supply to the population of Azerbaijan at the present time is urgent and requires the adoption of appropriate measures to solve it.

Selenium is an essential, absolutely essential element for the life activity of many organisms (from viruses to mammals) and, mainly, humans. Despite the fact that its gross content in a 70 kg human body is only 14–15 mg, it is directly involved in many vital regulatory processes [1–3]. Its distribution in the Earth's crust is insignificant, the so-called clark makes only 10 5%, and, thus, it is distributed very unevenly. It is accepted to consider soils that a content of less than 10 5% of selenium as poor and more than 10 5% as rich soils [4]. Proceeding from this the content of selenium in products depends on its regional provisioning and, consequently, provisioning of selenium (selenium status) in human organisms can vary greatly even within one country. At the same time, it was found that different organisms absorb selenium unevenly. Some plants belonging to cereals and astragals can serve as indicators of soil selenium supply.

Despite the fact that the selenium content in the ocean is very low, some species of aquatic organisms, including various algae (e.g., spirulina) have the ability to accumulate it in their tissues [4]. In addition to species specificity, there is also organ specificity. In the liver, kidneys, retina, thyroid gland, adrenal glands, testes, blood cells (lymphocytes, platelets, red blood cells), and nerve cells the selenium content is high, which indicates its importance in their functioning [3, 5].

However, despite the tremendous progress in the understanding of the biological role of selenium achieved over the last 50 years, its true potential as a biologically active substance is far from being disclosed. The history of research on the biological properties of selenium covers characteristic stages since 1817, from the moment of its discovery by I. Berzelius as a chemical element [6, 7]. In 1957 the American scientist K. Schwarz proved its anti-necrotic value in a number of animals, the so-called anti-necrotic factor - 3 [8]. Since then, the attitude towards selenium as a purely toxic element shifted to the desire to study its useful biological functions [9, 10]. Thus, in 1973 it was found that the previously well-known anti-peroxide, hemoglobin-protective enzyme glutathione peroxidase (*GPX*) [11] is a selenium-dependent protein, and its functions as an antioxidant are much broader than had been commonly thought [12, 13]. In 1970–1980 the existence of other selenoproteins was established, and that selenium is localized actually in all cells of the organism [14–17].

In the 1990s, three selenium-containing enzymes at different levels involved in regulating iodine metabolism were identified [18]. These discoveries stimulated even greater interest in its intracellular regulatory functions. Over 30 selenium-containing proteins have been identified in cells of various organs and tissues encoded by about 25 genes. Specific physiological functions were established in some of these proteins, while many of them had antioxidant properties [19].

At the same time, a unique mechanism of selenoprotein synthesis was discovered with the use of the so-called *SESIS* mechanism. It contains the 21 obligatory amino acid *Se*-cysteine (*Sec*), encoded by the *UGA* stop codon in the *mRNA* structure. Selenium is incorporated into selenoproteins via *Se* cysteinyl tRNA, which in turn is synthesized by transferring the selenium group into selenium-tRNA from selenophosphate. This mechanism is unique in that it is co-translational in that protein synthesis on ribosomes occurs simultaneously with the synthesis of the 21st amino acid (i.e., conversion of serine to *Se cysteine*) [19–24].

Farhan Saeed et al. show that there is great potential for selenium to affect the immune system, for example, the antioxidant peroxidase GSH probably protects neutrophils from oxygen radicals that are produced to destroy ingested foreign organisms [25]. Selenium affects both the innate, "maladaptive" and the acquired, "adaptive" immune system. Selenium-deficient lymphocytes are less able to proliferate in response to mitogen, and in macrophages, its deficiency impairs the synthesis of leukotriene B4, which is essential for neutrophil chemotaxis. The humoral system is also affected by selenium deficiency; for example, *IgM*, *IgG*, and *IgA* titers are reduced in rats, and *IgG* and *IgM* titers are reduced in humans [2].

Linda Johansson et al. showed that selenocysteine (*Sec*), the 21st amino acid, exists in nature in all kingdoms of life as the defining element of selenoproteins. *Sec* is an analog of cysteine (*Cys*) with a selenium-containing selenium group instead of the sulfur-containing thiol group in *Cys*. The selenium atom gives *Sec* completely different properties than *Cys*. The most obvious difference is the lower *pKa* of *Sec* and the fact that *Sec* is a stronger nucleophile than *Cys*. Proteins containing Sec are often enzymes that utilize the reactivity of the Sec residue in the catalytic cycle. Therefore, *Sec* is usually necessary for their catalytic efficiency [26].

Moghadaszadeh B. and Beggs A.H. in their article show an overview of human selenoprotein expression and function and schematically depict the process of *Sec* codon recognition and *Sec* insertion requiring several trans-acting factors including $tRNA^{Sec}$, *Sec*-specific elongation factor and *SECIS*-binding proteins. It has been observed that targeted deletion of the $tRNA^{Sec}$ *Trsp* gene leads to an embryonic lethal phenotype in mice [27]. To illustrate the scheme, let us show the above-described processes in **Figure 1**.

Thus, according to the author [27], all animal specific (acting) *Se* proteins are Se cysteine-containing natural compounds in the active center. In the organic world, selenium is usually in the form of the amino acids selenocysteine (*Sec*) and selenomethionine (*SeMet*), which differ in the presence of selenium instead of sulfur. This substitution is predictably related to the fact that selenium is closer to serine than to other chalcogenes in its physical and chemical properties: atomic radius value, electronegativity value and polarizability of the oxidation degree. All these parameters determine the increased nucleophilicity, which provides higher catalytic activity of *Se*-proteins in relation to their sulfur-containing counterparts. However, despite the obvious advances in this field, there is still no clear understanding of all sides of this mechanism.

The main differences in the chemical properties of selenoproteins and their sulfur analogues are due to a significant difference in the values of the dissociation constants (*pKa*), which for *Sec* is 5.1, and for *Cys* 8.3 [28, 29]. This circumstance makes thiolates (ionized form) less reactive than selenolates.

Kohrle J. reports that in experimental animal models prolonged and severe selenium deficiency leads to necrosis and fibrosis after high iodide loads. Combined iodide and selenium deficiency, such as in central Zaire, is thought to cause a

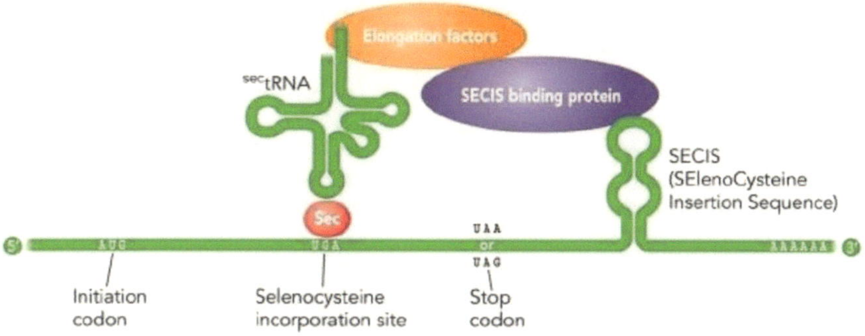

Figure 1.
Selenoproteins and their impact on human health through diverse physiological pathways [27].

myxedematous form of endemic cretinism. Insufficient selenium intake and diagnostically low serum selenium levels correlate significantly with the development of thyroid carcinoma and other tumors. Although selenium intake controls the expression and translation of selenocysteine-containing proteins, no direct correlation has been found between tissue selenium content and the expression of various thyroid selenoproteins, suggesting that other regulatory factors contribute to or override selenium-dependent expression control, such as in adenoma, carcinoma or autoimmune thyroid disease. Because both micronutrients, iodine and selenium, were leached from the topsoil during and after the ice age in many regions of the world, an adequate supply of these essential compounds must be provided by either a balanced diet or supplements [30].

Gustin C. et al. state that jodine (J) and selenium (Se) are necessary for the synthesis of thyroid hormones. Iodine and selenium interact. Pregnancy increases the mother's need for iodine [31]. And Mayunga K.C. et al. reported inadequate iodine levels in pregnant Dutch women [32]. Because as there is no enough information about their selenium intake, we examined iodine status and selenium intake in relation to iodine and selenium supplementation during pregnancy. The authors concluded that research on the 21st amino acid, selenocysteine, has progressed over the past 30 years from the intriguing discovery of *Sec* in a few select proteins to the recognition of *Sec* as an important component of many living organisms, associated with human disease and translated into an extension of the genetic code. The field of study of proteins naturally containing selenocysteine is growing rapidly, with new selenoproteins being discovered that have yet to be characterized. The ability to produce synthetic selenoproteins should facilitate such research, as well as open up new possibilities for biotechnological techniques based on the unique properties of selenocysteine. They are confident that the biochemistry of selenium-based proteins will form the basis for several future technologies of both fundamental and medical importance.

In experimental animal models, long-term and strong selenium deficiency leads to necrosis and fibrosis after high iodide loads. Combined iodide and selenium deficiency, such as in central Zaire, is thought to cause the myxedematous form of endemic cretinism [33]. The trace element selenium is of essential importance for the synthesis of a set of redox active proteins. Kamil Demircan et al. [34], studied three additional biomarkers of serum selenium status in relation to overall survival and recurrence after diagnosis of primary invasive breast cancer in a large prospective cohort study. They concluded that the prediction of mortality based on all three biomarkers was superior to established tumor characteristics such as histologic grade, number of lymph nodes involved, or tumor size. Se-status assessment at breast cancer diagnosis identifies patients at exceptionally high risk for poor prognosis.

2. The experimental part of the study

2.1 The main differences in the chemical properties of selenium proteins and their sulfur analogues

The main differences in the chemical properties of selenoproteins and their sulfur analogues are caused by a significant difference in the dissociation constants (pKa) values, which is 5.1 for selenocysteine (*Sec*), and 8.3 for cysteine (*Cys*) [28, 29]. This circumstance makes thiolates (in ionized form) less reactive than selenolates.

2.1.1 Se-proteins and their functions

Animal (mammalian) *Se* proteins are commonly divided into 3 categories: [3, 18].

1. True selenium-specific selenoproteins that include *Se* cysteine in the active center;

2. Proteins that do not include specific selenium; Selenium-binding proteins, such as (SBP 1)

Among the identified 30 specific Se proteins encoded by 25 genes, only a small fraction of them has specific physiological functions [19]. A hierarchy of "sensitivity" of *Se* protein synthesis to dietary intake of selenium has now been discovered and it is postulated that the hierarchy of *mRNA* expression is closely related (deterministic) to the importance of this or that selenoprotein in cellular hemostasis [35]. The organ-tissue specificity of selenoprotein distribution, i.e., their localization by tissue principle, exemplified by the glutathione peroxidase family, has also been established. While *GPX* is present in many cell types, *GPX* is expressed only in the gastrointestinal tract, *GPX* in intercellular medium and blood plasma, *GPX* in the nasopharyngeal epithelium, *TRXR*3 is localized in testes, iodothyronine deiodinases in thyroid tissues, etc. [36].

The high antioxidant properties of selenium were first established back in the 60-the 70s of the last century. And since the previously well known antioxidant enzyme (*GPX*) turned out to be a selenium protein, the *AO* properties of many newly discovered *Se* proteins were discovered. However, *Se* proteins were found to have many other important biological properties in addition to their antioxidant properties, such as regulation of thyroid hormone activity, participation in the regulation of non-specific immune response, inhibition of inflammatory, chemotactic, and phagocytic reactions, influence on reproductive functions (male infertility), participation in redox reactions. The authors [36] briefly describe both the function of these selenoproteins and the regulation of their expression depending on *Se* status and tabulate data for 40 proteins important for understanding the function and significance, effects of dietary selenium, and subcellular localization.

2.1.2 Se accumulating proteins

It turned out that UGA serves as a stop-signal and selenocysteine codon in the genetic code, but there are no computational methods to determine its coding function, which means that most selenoprotein genes are wrong. Gregory V. Krukov et al. identified selenoprotein genes in sequenced mammalian genomes using methods based on determining structures of selenocysteine *RNA* insertions by coding for *UGA* codon potential and presence of cysteine-containing homologs. They found that the human selenoproteome consists of 25 selenoproteins [37].

Based on the *SECIS* method applied to mammalian genomes, the authors identified *SECIS* candidate elements in the human genome using the *SE CIS*2. 0 program [37]. Structural and thermodynamic features of *SECIS* elements were analyzed using this program. The candidate elements were about 10 times more selective (for the same specificity) than the original SECISearch version [38]. They then identified human/mouse and human/rat *SECIS* pairs using the SECISblastn program, which analyzes the evolutionary conservation of mammalian *SECIS* elements. In addition, they analyzed genomic sequences upstream of *SECIS* candidate elements using geneid [39], a gene prediction program that identifies open reading frames (*ORFs*) with high coding potential and containing infra-labeled *TGA* codons.

By analyzing predicted human selenoprotein genes using *MSGS* (mammalian selenoprotein gene signature) criteria [37, 40], which test selenoprotein homologs for the presence and conservation of *ORFs* intraframe *TGA* codons and *SECIS* elements, the authors concluded that *SelH*, *SelI*, *SelO*, *SelS* and *SelK mRNA* are present in various tissues and cell types. However, *GPx6mRNA* was found only in embryos and olfactory epithelium, and *SelV mRNA* expression was limited to the testes, where it was present in the seminal tubules. The authors' predictions regarding the secondary structure and organization of the protein showed that, like all previously described mammalian selenoproteins, *GPx6*, *SelH*, *SelO*, and *SelV* are globular proteins. However, *SelK* and *SelS* were predicted membrane proteins. They expressed *SelK* and *SelS* fusions containing the *C terminal* tag of green fluorescent protein (*GFP*) in *CV* 1 cells and found that the fusion products were indeed on the plasma membrane. Thus, *SelK* and *SelS* appeared to be the first known selenoproteins of the plasma membrane.

SBP selenium-binding proteins can be said to be included proteins in which the form of selenium is unknown. Although *Se* is stably bound, probably through the selenosulfide bond. One of them, *SeBP* 1 (*Se* Binding protein), has been intensively studied recently due to its prominent role in tumor growth [41, 42].

2.1.3 Relationship of Se and hemoglobin

The comparative distribution of *Se* over the two major erythrocyte proteins, *HA* and *Hb*, in humans and animals with different selenium metabolism (different sensitivity to *Se* deficiency) was studied in detail in 80 90 years by such researchers as M.A. Belstein, J.A. Butler, K.D. Thomson, P.D. Wanger and others [43]. They showed the predominance of *Se* inclusion in human and some primate hemoglobin (90% of all *Se* in erythrocytes) versus low *Se HPC* coverage (10%). At the same time, in the erythrocytes of animals sensitive to *Se* deficiency, such as sheep, rats, hamsters, etc., the proportion of *Se* included in the *HPC* is significantly higher than in humans, some primates, etc. These objects in conditions of selenium deficiency signs of sensitivity of selenium deficiency pathologies (liver and kidney necrosis, white muscle disease, exudative diathesis) and have rather high levels of *GPX* activity in organs and in erythrocytes, and their hemoglobin has a low capacity (0.1 0.2) to absorb selenium. Organisms (guinea pig, human, some primates) sensitivity dependence on selenium deficiency usually also have reduced *GPX* in the organ activity, and most of the intraerythrocyte selenium is included in the hemoglobin fraction.

Using the example of the inhabitants of Azerbaijan (Baku), we have shown that 3/4 of erythrocyte *Se* enters the hemoglobin fraction at a ratio of 1 *Se* atom per 300–1000 *Hb* molecules. Selenium is incorporated into hemoglobin by sulfur substitution predominantly in cysteine residues at the *βCys*93 position. Considering that it will affect the electronic environment of proximal histidine, which is in close proximity to heme, one can assume that it will enhance its antioxidant protection [43–45].

We examined the effects of sodium nitrite and sodium selenite in their joint and single action on the processes of oxidation of hemoglobin (*Hb*), lipid peroxidation (*LPO*), the activity of antioxidant (*AO*) enzymes glutathione peroxidase (*GP*) and catalase in human red blood cells *in-vitro*. Nitrite was found to have a significant effect on the oxidative processes in erythrocytes and *Hb*, while sodium selenite attenuated the development of the nitrite-induced oxidative process in erythrocytes and reduced the formation of methemoglobin (*MetHb*) by 25–40%. Having a significant effect on the oxidative process in erythrocytes, nitrite does not lead to a marked increase in lipid peroxidation rates in erythrocytes. Under the influence of nitrite, there is a slight change in the activity of *AO* enzyme *GP* (up to 20–30%), and the activity of catalase in all cases drops significantly (1.5–2 times). Nitrite in the incubation medium increases the concentrations of membrane oxyhemoglobin and *MetHb*, while sodium selenite has an inhibitory effect on this process [46–48].

Based on the fact that in the human body de novo synthesis occurs for a long time (up to 48–72 hours) in the liver and in the ready form comes with the blood stream to the erythrocytes, experiments were conducted to study the oxidative resistance of erythrocytes and hemoglobin to the damaging effects of such environmental factors as high pressure electric field, ozone, UV-radiation [43]. Here it was found that selenium incorporated into hemoglobin during the first 2 hours increases resistance to them without additional contribution of *AO* selenium-induced synthesis of *GPC* enzyme. On the other hand, it was shown that under conditions of selenium deficiency (blood of pregnant women, as a natural model of selenium deficiency) hemoglobin is impoverished in selenium, as are red blood cells, which is accompanied by a decrease in the antioxidant properties of *Hb* and red blood cells.

At the same time, the *Hb* activity in erythrocytes is weakly altered even in the third trimester of pregnancy. This is further evidence that *Hb* enzyme activity does not always adequately reflect selenium status [43, 44]. Regarding the effect of selenium on the health of pregnant women, it can be noted that pregnancy pathologies such as threatened termination, intrauterine fetal delay are accompanied by a decrease in selenium levels and *Hb* activity in serum, erythrocytes with an increase in lipid peroxidation (*LPO*) of erythrocytes [43, 44]. Selenium deficiency has been found to impair the regulation of nutrient transport through the placenta [49, 50]. In addition, serum selenium levels may serve as a risk marker for hypertension in pregnancy [51]. In addition, we can add that selenium deficiency can affect many health parameters, including the cognitive functions of children in the first few years of life, and also significantly increases the risk of adverse pregnancy development in various infections [52, 53]. The effect of sodium selenite on the development of lipid peroxidation (*LPO*) was studied. We also studied the accumulation of methemoglobin (*MetHb*) by selenium, the state of reduced glutathione (*GSH*) and glutathione peroxidase (*GP*) activity in isolated erythrocytes in incubation medium containing different final concentrations of sodium selenite (Na_2SeO_3). Low (1 M, 5 M) concentrations of sodium selenite were found to have little effect on glutathione, while at high (50 M and 100 M) concentrations there was a marked depletion of glutathione, and the activity of glutathione, which has glutathione as the main oxidation substrate, was also significantly reduced. Characteristically, high-end concentrations of lead to increased oxidative processes in both hemoglobin and erythrocytes. Conversely, low sodium selenite concentrations lead to a decrease in the accumulation of active thiobarbituric acid (*TBA*) and *MetHb* products. It has been suggested that the stimulation of oxidative processes by high concentrations of sodium selenite is associated with the inhibition of the key antioxidant enzyme *GP*, which is due to the formation of *Se* [48].

2.1.4 Selenium in gerontology

Aging can be represented as a process of continuous destruction inherent in all objects of animate and inanimate nature, a consequence of the second principle of thermodynamics, and an organism as an open thermodynamic system that dissipates its heat and simultaneously consumes free energy of high-potential light or chemical from outside. The existence and maintenance of complex dissipative structures of living organisms is possible due to the constant flow of energy, as well as the continuous reproduction of genetic information and structures in the process of cell division. Agerelated changes in somatic cells of multicellular organisms are caused by a decrease in proliferative potential and free radical reactions, the main source of which is oxygen reduction performed by mitochondria, microsomes, and *NADPH* oxidant systems of phagocytes and other specialized cells.

According to V.A. Gusev, the magnitude of the flux of reactive oxygen species is related to the intensity of the basic metabolism. The accumulation of damage in cells and the rate of aging depend on the ratio of reactive oxygen species formation and their deactivation by the enzymatic antioxidant defense system. The reason for the inevitable occurrence, leakage and dissipation of reactive oxygen species during energy conversion in mitochondria is the second law of thermodynamics, which excludes 100% efficiency of such processes. Comparison of specific superoxide dismutase activity in human granulocytes, platelets, erythrocytes and lymphocytes with the ability of these cells to exogenously generate superoxide radicals allowed to trace the relationship of these factors to the lifetime of cells in blood, which varies from 12 hours to several years [54].

A physiological process, similar to pregnancy, associated with the weakening of AR status and activation of free-radical processes is old age. Currently, there are two main hypotheses of the development of old age, one of which is genetic, i.e. programmed, and the second one is based on the acceleration of free-radical processes leading to *AR* depletion in the organism [35]. This hypothesis was first proposed by Harman D. and is still a priority [55]. Although there is no clear link between these hypotheses, there is strong evidence that free-radical reactions accelerate with age, having a negative impact on physiological processes related to age [56]. *AO* minerals such as selenium and zinc have been found to be involved in maintaining metabolic homeostasis in older adults.

Their deficiency increases with age, which is probably a significant cause of premature aging [35]. H. Steinbrenner and S. Helmut [57], believe that antioxidant selenium enzymes as well as pro-oxidant effects of selenium compounds on tumor cells are involved in the anticancer effects of selenium. Brigelius-R. Floĥe and M. Matilde [58] argue that collectively, selenium-containing *GPx* (*GPx*1, *x*4&*x*6) as well as their non-selenium congeners (*GPx*5, *x*7&*x*8) have become key players in important biological contexts far beyond hydroperoxide detoxification. In the pathogenetic mechanisms of aging, *LPO* activation plays an important role against the background of decreased *AR* status of the organism, which can be corrected by the use of *Se* drugs.

Using the example of the inhabitants of Azerbaijan (Baku), we have shown that 3/4 of erythrocyte *Se* enters the hemoglobin fraction at a ratio of 1 *Se* atom per 300–1000 *Hb* molecules. Selenium is incorporated into hemoglobin by sulfur substitution predominantly in cysteine residues at the *βCys*93 position. Considering that it will affect the electronic environment of proximal histidine, which is in close proximity to heme, one can assume that it will enhance its antioxidant protection.

2.1.5 Selenium and iodine deficiency conditions

In the development of iodine deficiency states, in addition to iodine itself, as it has been discovered relatively recently, in the last 20–25 years, the provision of the trace element selenium to the body is of great importance. This is the main molecular synergist that has key regulatory significance in thyroid gland (*TG*) functioning. Characteristically, iodine and selenium act at the cellular level in all organs of the body, with amounts and requirements of the same order (14 mg (Se) and 20–35 mg (*J*)), and daily intake is (60–120 mg *Se* and 150–250 mg *J* [59, 60]). It turned out that many patients have a clear selenium deficiency along with iodine deficiency, indicating that iodine deficiency conditions (including goiter) cannot be cured by iodine supplementation alone. It has been experimentally proven that even under conditions of normal iodine intake, selenium deficiency leads to necrosis and thyroid fibrosis [61]. The importance of not only iodine, but also selenium in the treatment and prevention of thyroid diseases is recognized by all leading specialists, and the study of this problem is urgent [62].

It is now established that selenium is involved in the metabolism of thyroid hormones because it is a component of deiodinases, a family of selenoenzymes including selenocysteine and 5′-iodothyronine involved in the transformation (conversion) of *T* 4 to *TK*, performing deiodination of the outer ring of *T* 4. Deiodinases belong to the family of selenoenzymes that include selenocysteine. One of the important enzymes responsible for the conversion of thyroxine to 3, 5, 3′*triiodothyronine*, 5′ *iodothyronine* deiodinase type 1 (*D*1) [18, 63], was first shown to be a selenoenzyme in 1990–1991. The findings explained why the conversion of *T* 4 to *TK* was reduced in the selenium-deficient experiment, leading to the development of hypothyroidism symptoms. Many studies have focused on deiodinase type 2 (*D*2). In humans, plasma *T* 3 is formed in the thyroid gland (20%) and by peripheral deiodination (80%).

Accordingly, the role of *D*1 and *D*2 in the formation of circulating *T* 3 remains unknown, but there is speculation that *D*2 may play a greater role in this process. Deiodinase type 3 (*D*3) catalyzes the conversion of *T* 4 and *T* 3 to inactive metabolites [64]. It is expressed in high concentration in the placenta and regulates the concentration of circulating fetal thyroid hormones throughout gestation. The action of selenium-dependent deiodinases in tissues is under the control of the selenium diet and is realized with the participation of thyrotropic hormone [65, 66]. The effect of both isolated selenium deficiency and selenium deficiency combined with iodine deficiency on the human body is of interest to researchers, since pronounced combined deficiency of these elements is a problem in many regions of Central Africa (Congo, Zaire, Sudan), Tibet and some European countries [62].

Of particular interest is the fact that during pregnancy iodine deficiency often leads to the development of thyroid diseases, mainly due to the doubled need for iodine and other important elements, primarily selenium, the lack of which in addition to its direct effect on iodine metabolism and thyroid hormones contributes to other dangerous pathologies, including infant mortality syndrome [67]. It should be added that all over the world due to deteriorating environmental conditions (heavy metals, acid rain, intensive chemicalization of agriculture, etc.) the content of mobile forms of selenium in soils is constantly decreasing, which is reflected in the selenium status in the human body. The role of selenium in the development of iodine deficiency states is not fully understood, and data on the relationship between selenium deficiency in food and preservation of thyroid function require further study [62].

Taking into consideration that deficiency of iodine and selenium in living organisms increases the risk of thyroid gland diseases, malignant neoplasms, cardiovascular pathology, and other serious diseases, the issue of provision of an organism with these microelements is actually all over the world, including *CIS* countries. This problem is also extremely important for Azerbaijan. It is noted that the microelement selenium is closely connected with iodine metabolism in organisms that is of key importance for thyroid gland functioning. The importance of not only iodine but also selenium in the treatment and prevention of thyroid diseases is recognized by all leading world experts studying this problem. In this connection it is necessary to further study in detail the joint functioning of these elements in organisms, consider the development of a new state strategy for the liquidation of iodine deficiency in Azerbaijan, and possible revision of current salt iodization program in favor of the medicinal prophylaxis with iodine-containing oil capsules with additional use of selenium preparations and continuous monitoring of iodine supply, use the existing positive experience of the international organization "World Doctors" (1998–2004) [62].

2.1.6 Se and immunity

A number of micronutrients, including *Se*, are known to be important in maintaining a "proper" immune response. Selenium is essential for the efficient formation and functioning of virtually all components of the immune system, including the major immune cells: neutrophils and macrophages, *NK* (natural cell) killers, *T lymphocytes* and *B lymphocytes* [68, 69]. In particular, it is well known that high *Se* levels in the body stimulate the proliferation and differentiation of *CD*4 + *T* helper cells (*Th*) [70]. Selenium is also important for the cytotoxicity function of *CD*8 + *T* cells and *NK* cells *Se* levels have a significant impact on innate immunity function, in particular macrophage activity depends on selenium levels for their signaling and antigenic abilities [69]. Added to this is the fact that selenium is actively involved in regulating the activity of such interleukins as *IL* 1, *IL* 6, *IL* 10, *TNF* through the coordination of the nuclear transcription factor *NF kB*, which is inhibited by selenium. At the same time, the expression of such inflammatory cytokines as *IL* 2, *IL* 8, and *IL* 18 is stimulated [71].

T cells have an increased sensitivity to oxidative stress, and when deficient in selenium proteins, *T cells* cannot proliferate in response to stimulation of T cell receptors due to loss of generation of reactive oxygen species and nitrogen [69, 70].

To date, the *Se* proteins involved in the formation of the immune response have been most fully characterized: the *GPX*, *TXNRD*, and *DIO* families and proteins such as *MSRBI*, *SPS*2 [69].

Analysis of the available data suggests the effect of selenium deficiency on innate and adaptive immunity. However, selenium supplementation does not always produce positive results. This is particularly evident in the case of tumor growth, where there are no clear positive results on the use of selenium supplementation for cancer control [72].

We will not address this topic in detail in our review, but we will note the main points. Back in the 60's and 70's a Canadian researcher R. Schamberger noted that in biogeochemical provinces rich in selenium the incidence of cancer was much lower than in selenium-poor regions [73]. This work initiated a broad study of the role of selenium in tumor growth. In the 70s on the initiative of Prof. G.B.Abdullaev our laboratory staff began to study migration of endogenous and exogenous selenium in the rat organism - Giren carcinoma, Wakor carcisarcoma, M1 sarcoma. A sitadic character of selenium accumulation in these tumors was shown (exchange of selenium between the tumor and rat organs and tissues), i.e. affinity of selenium accumulation in malignant tumors was established, which suggests that tumors need selenium as an antioxidant for their development [74]. Established on experimental animals inhibition of tumor growth by a number of selenium compounds stimulated their use as adjuvants in oncology. However, conflicting results were obtained here [72].

We studied the sitadic nature of selenium accumulation in these tumors (exchange of selenium between tumor and rat tissue organs). We found that selenium atoms accumulate affinely in malignant tumors. This suggests that tumors need selenium - as an antioxidant - for their development. And a high dose destroys them. This was reported at the 1st Scientific Conference "Selenium in Biology" in Baku, 1974 [74].

In this regard, some researchers have tried to use already toxic doses of selenium compounds to apply them as proxidants, which can penetrate into tumors as toxicants and thus inhibit tumor development. In some cases, positive results are achieved on esperiments, but this is not universal. Therefore, manipulations of individual selenoproteins at sub-toxic doses may be useful to study the immune system and to identify the molecular mechanisms of selenoprotein regulation of immunity. These mechanisms should include pro-oxidative and proteomic activities that provide suppression of cancer development (apoptosis, necrosis, paranthosis) [72].

2.1.7 Selenium as an antitoxicant in nitrite poisoning

One of the main targets of the toxic effects of nitrites is hemoglobin, which has an increased oxidative affinity (formation of methemoglobin and other oxidative derivatives) for nitrites [75]. There is extensive data on the use of antioxidants of different nature to attenuate nitrite toxicity, including through the break-down of nitrite metabolites (peroxynitrite, etc.). In particular, there is data on the AO action of selenium-containing substances: *Se*-proteins and *Se*-amino acids or other selenium compounds (usually acting similarly to *SH*-containing compounds, but with a greater efficiency) [76, 77]. There is evidence that some selenoproteins can catalyze the breakdown of *ONOO* (an aggressive radical capable of oxidizing cellular structures) with a high 2nd order final reaction rate. It has been suggested that *hP x* acts as a peroxynitrireductase, reducing *ONOO* and protecting hemoglobin from oxidation and nitrification [78].

There are several indications in the literature that sodium selenite is readily incorporated into erythrocytes (selenium pump), where it undergoes complex metabolism, interacting with hemoglobin, affecting its properties, with subsequent release from erythrocytes into plasma as part of various albumin [79, 80]. Thus, selenium incorporated into erythrocytes as an active intermediate can affect oxidative processes induced by nitrites or their metabolites. The transfer of selenium from erythrocytes into plasma is carried out through the membrane anion exchanger *AE*1 through a complex interaction of membrane *SH*-proteins including transported selenium, interaction with plasma albumin. *NO in vitro/in vivo* is formed through the inherent nitrite reductase activity of hemoglobin according to the scheme: $Hb + NO^2\text{–}MetHb + NO + H_2O$ [81].

On the other hand, *NO*, as the main metabolite of the NO^2– ion in vitro and in vivo, interacts with hemoglobin in the same complex way, binding directly to heme (nitrosyl hemoglobin *HbNO*) or including in the *SH* group of α or β *peptide* chains (nitrosohemoglobin *SNHb*) as NO^+ nitrosonium cations [48, 75]. Of particular interest is the incorporation of *NO* into the β *chain* of hemoglobin at the β*Cys*93 position, which has important physiological significance for its vasodilator function. This circumstance is also interesting because selenium from sodium selenite, i.e., selenium replacing sulfur in the β – *chain* of cysteine, is also included in this position. In other words, selenium, along with *NO*, is included in the same site of the hemoglobin β *chain* (β*Cys*93) [79, 80].

At the same time, the frequency of selenium presence in *Hb* for humans in norm according to one data is 1: 225 [80], according to other data *Se*: *Hb*1: 300 [43, 44].

Normally, the frequency of *NO* inclusion in *Hb* is *NO*: *Hb* 1: 1000 (but in extreme cases may reach 1: 100), i.e., the number of inclusions in *β chain* is normally higher for selenium than for *NO*, and the inclusion of *NO* directly in *β chain* is even lower (40%) [82].

When dietary conditions change (nitrite poisoning or nitrogen deficiency) of both nitrite and selenium (excess or deficiency in the diet), the *NO*: *Hb*1: 1000 and *Se*: *Hb* 1: 300 ratio may change significantly, especially for nitric oxide due to the extensive use of nitrate/nitrite in agricultural production and food industries. In this case, excess *NO* can stimulate oxidative stress as one manifestation of nitrite toxicity. Thus, inclusion of selenium at the same site (*βCys*93) may create competition for *NO* and thereby reduce the oxidative burden on hemoglobin, in addition to the action of *GPx* as a natural defender against oxidation.

Moreover, relatively recently, it was shown using transgenic mice that the amino acid residue *β* 93*sus* itself confers certain *AR* properties on erythrocytes during hydrogen peroxide stimulation of the ferric forms of hemoglobin [83]. Earlier, a similar idea was put forward by Mansouri [84] when studying the sodium-dependent oxidation of hemoglobin, that *βCys*93 has a protective *AR* function for hemoglobin. As for selenium, we previously showed that a 2-hour incubation of human erythrocytes with sodium selenite (Na_2SO_3) leads to a doubling of the selenium content in the hemoglobin fraction, increasing the *AR* properties of both hemoglobin and erythrocytes (*LPO* reduction). The authors explain this by the lower electronegativity of selenium atoms in relation to the sulfur atoms they replace [48].

The question of how such low NO inclusions in hemoglobin can exert significant physiological effects remains to be fully elucidated, despite impressive achievements in this field (recognition of *NO* as a gas molecule, etc.). To a certain extent, this also applies to selenium, whose content in hemoglobin is comparable to *NO*, but its physiological role, in addition to that of *AR*, has not been elucidated. And the fact that an essential part of *NO* in hemoglobin is at the same site together with selenium suggests a close interaction of these two ligands in comparable proportions. Which makes it interesting to study this issue.

2.1.8 Selenium regulation of oxidative processes in blood of rats induced by sodium nitrite

The role of selenium in moderate doses of sodium nitrite on rat erythrocytes was studied in vivo. Rats were exposed to single and combined Na_2SeO_3 [0.5 mg/kg] and $NaNO_2$ [30 mg/kg] by intraperitoneal injections and subsequent exposures with periods of 1, 2, 3, and 12, 48 h. Administration of sodium nitrite with exposures at 1 and 3 h in rats resulted in a marked accumulation of *MetHb* and already by 1 h reached 30%, which during the following 2–3 h monotonically decreased to 30% of the maximum level reached. By 12 and 48 h of exposure, the level of *MetHb* was little or no different from the control, respectively. Under the action of nitrite in the erythrocyte suspension was found to decrease (by 30% of control) the content of products reacting with thiobarbituric acid (*TBA*). A single injection of sodium selenite did not lead to changes in *MetHb* and lipid peroxidation (*LPO*). At short-term exposure (1–3 h), combined administration of selenite and sodium nitrite resulted in a decrease in nitrite-induced accumulation of MetHb by 35% and an increase in accumulation of *LPO* products compared with the single nitrite action. At the same time, the order of administration had no effect on the final result.

At prolonged exposure, preinjected selenite at 48 h followed by nitrite [with 1 h incubation] led to a decrease in nitrite-induced MetHb accumulation by 16 and 41%

of *LPO* values, whereas selenite injected 1 h after nitrite [48 h exposure] had no effect on *MetHb* accumulation and slightly (10%) reduced *LPO* values. Changes in the activity of antioxidant enzymes, glutathione peroxidase, and catalase, were examined. The activity of catalase decreased in all variants of exposure to sodium nitrite. Selenite did not lead to a significant increase in the activity of *GPX* under short-term exposure, while nitrite led to its inhibition. Exposure to selenite combined with nitrite had little effect on the $NaNO_2$-induced decrease in *GP* activity. The decrease in nitrite-induced accumulation of *MetHb*, when sodium selenite is administered during the first 1–3 h, is probably more related to the very fact of selenium inclusion in the *Hb* molecule than to the effect of additional contribution of *GP*, whose activity is not significantly increased during this period of exposure. Based on the position of the spectral maxima for HbO_2 and *doxHb*, we note that $NaNO_2$ increases *MetHb* by reducing HbO_2, and selenite inhibits this effect [47].

2.1.9 Se and Covid-19

The discovery of a significant role of selenium deficiency in COVID-19 development has led to increased interest in the question of selenium-virus interactions. To date, there are many studies on this topic, a huge amount of clinical material has been accumulated, but a number of unresolved questions remain.

Here we will touch upon only some of the issues in the interaction of selenium with viruses in humans [85–87]. The mechanism of selenium antiviral action is multifaceted and covers a number of stages of viral infection, from virus invasion into healthy cells to fighting its consequences. Below is a brief list of the beneficial properties of selenium sodium selenite (the main inorganic selenium compound used in biology and medicine) in the treatment of viral infections, using *HIV* and *Ebola* as examples [85–87]. Sodium selenite (Na_2SO_3) can act as a contact interrupter between virions (*SARS CoV* 1, *SARS CoV* 2) and the membrane apparatus of healthy cells (*host*). Specifically, the *SARS CoV* 2 virion itself consists of a hydrophobic envelope with protein spikes on the outside and a carrier of its genome, *mRNA*, on the inside.

The proteins of these spikes interact with the membrane apparatus of the "*host*" cells, i.e. the organism attacked by the virus, mainly through the membrane integral cell protein, the angiotensin-converting enzyme *ACE*2 (angiotensin) and with the subsequent disruption of membrane integrity, facilitating the penetration of the virus genetic material into healthy cells. Subsequently, this *mRNA* is incorporated into the host cell genome, modifying it, after which the virus replicates at the expense of the host cell resources [88, 89]. Thus, interrupting the contact of virus spikes with the membranes of healthy cells by changing the structure of any spike proteins is a preventive measure to suppress the development of infection [90]. This hypothesis is presented in detail in the work of M. Kieliszek and B. Lipinski [91].

Sodium selenite (Na_2SO_3), being a small and non-polar molecule, easily passes through cell membranes by passive transport, has an active intracellular metabolism of selenium, which is accompanied by oxidation of intracellular sulfur-containing proteins with simultaneous reduction of selenite (+4) to selenide (2). Taking into consideration that selenium and sulfur are quite similar in their chemical properties, it can be supposed that when entering the body as a chemically more active element, selenium will replace sulfur in sulfur-containing cysteine (2 *amino* 3 *mercaptopropanoicacid*) or when interacting with *SH*-groups of proteins it takes away the hydrogen atom from thiols, thereby oxidizing them, forming *R S S R* and *R S Se S R* type bonds

[92, 93]. In the case of viral infection, sodium selenite will also interact with viral sulfur-containing proteins, including disulfidisomerase (*PDI*) located in *Covid* - 19 spikes, deactivating it as an enzyme according to the scheme:

$$PDI-(SH)_2+Se^{4+}\rightarrow PDI-S-S-PDI+Se^{2+} \tag{1}$$

This means that sodium selenite can contribute to the disruption of contact viral entry into healthy cells [90, 91]. As mentioned above, genomic antisense interactions lead to selenium deficiency, which leads to a decrease in selenium enzyme resources, primarily thyroredoxin reductase, a supplier of protons for the needs of DNA synthesis in healthy cells. This leads to increased consumption of selenium by the body, which is necessary for the synthesis of selenoproteins, both own and "viral". As a consequence, a selenium deficiency condition occurs, leading to the formation of reactive oxygen species [94], weakened immunity against the background of oxidative stress and decreased antioxidant protection of the body. Sodium selenite is a successful form of selenium in this respect, promoting its rapid penetration into cellular structures and overcoming the blood–brain barrier [95]. This property allows the body to use selenium from sodium selenite to maintain vital selenoprotein levels, protecting it from oxidative stress.

The main arguments for using sodium selenite in adjuvant treatment are as follows: 1. In model experiments, selenium inhibited *RNA* and *DNA* polymerase reactions; 2. Inhibited nuclear factor *NF kB* activity; 3. Regulates immune response, including inflammatory process; 4. it has an anti-aggregation effect by inhibiting the formation of thromboxane [86].

2.2 Preliminary research towards selenium-enriched protein - natural silk fibroin

Bioactive peptides are known for their high tissue affinity, specificity and effectiveness in health promotion. In this sense, fibroin and sericin of natural silk have a special place. Natural silk is a valuable textile raw material of animal origin. It is a product of excretion of silk-producing glands of animals, mainly silkworms (type of arthropods, class of insects). Among them, the most industrially important is the domesticated mulberry silkworm (*Bombyx mori L.*, a mulberry type silkworm), which feeds on mulberry leaves. By the end of V age, the caterpillars reach maturity and curl up into a cocoon that protects the pupa from adverse environmental conditions and silkworm enemies. Maturity occurs when a dense mass of silk, namely the protein fibroin (pure silk thread) and the protein sericin (sticky mass), is formed in the caterpillar silk gland.

If we consider the consumption of silk proteins, fibroin and sericin, from cocoons as bioactive peptides and hydrolysates of food proteins, which are known to be beneficial for human health, then modern silk production should contribute to food production and therefore equally to clothing, food and housing.

Enzymatic hydrolysis is a powerful tool for producing bioactive peptides and hydrolysates from fibroin and sericin. Motoyuki Sumida and Vallaya Sutthikhum [96], based on their experience of studying silk digestion enzyme for over 20 years, summarize current knowledge on bioactive peptides and hydrolysates produced from *B. mori L.* and wild silkworm fibroin and sericin using proteases, and their potential for human health promotion. They encourage researchers associated with silk proteins - fibroin and sericin - to conduct further comprehensive research on bioactive peptides and hydrolysates of fibroin and sericin derived from domesticated and wild silkworms. As such, these ingredients are expected to become fruitful resources for the well-being of mankind. In keeping with this principle, our results on fibroin enrichment with selenium are also becoming important in this field.

Furthermore, the aqueous solution of silk fibroin is suitable for preparing various silk fibroin films, hydrogels, porous materials, microspheres and the like used in cosmetics, skin care products, tanning lotions, tissue-engineered materials, drug carriers, artificial skin and the like. Since stable aqueous silk fibroin solution can be stored for a long time [97], it is obvious that enriching fibroin with selenium simultaneously increases the intelligence and innovativeness of aqueous fibroin solution.

Ch. Wen et al. [98] note that conventional inorganic *Se* supplements have drawbacks such as toxicity and low bioavailability. Enriched *Se* proteins and their hydrolysates show good bioactive properties, mainly including antioxidant activity, immune regulation, neuroprotective activity and inhibition of hyperglycaemia, among others. The authors advise that future studies should focus on the relationship between the metabolism of *Se*-enriched proteins and the metabolic pathways of selenoregulatory proteins using multiomics technology. In addition, in their opinion, the structure–activity relationship of *Se*-enriched proteins/hydrolysates from different sources should be comprehensively studied to further elucidate their bioactivity mechanism and test their beneficial properties *in vivo*. Considering this, as well as the findings of M. Puccinelli et al. [99] that increasing the amount of selenium in plant foods is a good way to increase *Se* intake in animals and humans, and the advice of the authors [96] above, our results on fibroin selenium enrichment may become important in this field.

2.2.1 Introduction of selenium into the fibroin structure

Selenium was introduced into the fibroin structure using our developed method [100]. Two batches of "Sheki-2" silkworm caterpillars were selected for this purpose. Starting from the fourth instar, the experimental batch was fed a preparation of sodium selenite (Na_2SeO_3); fresh mulberry leaves before feeding were sprayed with 0.1% solution of sodium selenite in distilled water, carefully dried, then caterpillars were fed every 48 hours. The dose of sodium selenite was taken at the rate of 4 mg per kg of live weight of the caterpillars. A control batch of caterpillars was fed with normal mulberry leaves. The temperature, humidity, light and feeding frequency of both batches were the same.

2.2.2 Preparation of pure fibroin

To purify fibroin obtained from silkworm (B. mori) cocoon filaments, we used the well-known sericin dissolution method [101]. Equal volumes of 0.05 M solutions of sodium carbonate Na_2CO_3 and sodium hydrogen carbonate $NaHCO_3$ were taken and the cocoons freed from their shells were boiled in them for 30 min. This allows fibroin to be separated from sericin. After washing fibroin five times in warm distilled water, the residual sericin in the sample was tested using a biurette reaction as follows: 2 ml of water remaining after the third wash of silk fibroin was added to a double volume of 30% $CuSO_4$ solution and the mixture was stirred again thoroughly. If sericin is present in the sample, it turns red-purple. Washing was continued until the sericin was completely absent.

The obtained fibroin was dried in a desiccator at 340 K, in glass cups, until constant weight. Fibroin was then extracted for 12 h with ethyl alcohol (20 g fibroin/500 ml

95% ethyl alcohol) to remove the waxes and for 12 h with petroleum ether (20 g fibroin/500 ml petroleum ether) in a Soscelet apparatus (extractor) to remove the fats.

2.2.3 Determination of selenium content in fibroin

The photometric method of selenium determination is one of the most convenient and up to now widely used in analyses of this element. This is primarily due to the availability of analytical equipment and the convenience and simplicity of the method [102]. To determine selenium content in fibroin, we used fluorimetric method adapted for biological samples [103]. Based on the ability of selenium to form in dilute solutions with 2, 3 - *diaminonaphthalene* a fluorescent complex - *diazoselenols* with a wide area (λ_{max} = 520 *nm*), when excited by UV light with λ_{max} = 366 *nm* (**Figure 2**).

The sensitivity of the method is 0.002 μg selenium per 1 ml of extract. Selenium content was determined in fibroin, its crystalline part and raw silk. Therefore, mineralization of the samples was carried out first. For this purpose a mass of dry sample (100 mg) was poured with concentrated nitric acid (5–7 ml), incubated for 24 hours in the dark, then 3–4 ml of 30% chloric acid was added. Using a reflux condenser the resulting mixture was heated first on a weak flame for 30 min and then on a strong flame.

A solution of $HClO_4$ was added from time to time and waited for the appearance of white vapors of perchloric acid until the solution was completely discolored. After cooling down, 10 mL distilled water was added to the mixture and heated again until the perchloric acid vapor appeared. Then the mixture was cooled down again and 2 mL of a 2% Determination of selenium content in fibroin *Ethylenediaminetetraaceticacid* (EDTA) solution was added. The *pH* of the solution was then adjusted to 1.0 using 10.0% concentrated hydrochloric acid and 25% ammonia solution. The mixture was stirred and 5 ml of 0.05% solution of 2, 3 *diaminonaphthalene* (in 0.1 N *HCl*) was added. The solutions were put on a boiling water bath for 5 min, cooled in the dark for 30–40 min. Then they were poured into a separating funnel with 5 ml of freshly distilled cyclohexane (or hexane) and extracted for 1 min. After separation of the phases the aqueous solution was discarded and the organic phase was poured into a cuvette for measurement. Fluorescence was measured on a sensitive FAS-1 fluorimeter. In each batch of determination a blank test was run through the whole assay cycle and an appropriate correction was introduced into the calculation of the selenium content of the samples. The selenium content of the test samples was calculated by plotting calibration curves.

In each batch of determinations a blank test was carried out throughout the analysis and an appropriate correction was entered into the calculation of the selenium content of the samples. By constructing calibration graphs, the selenium content of the test samples was calculated.

Daily measurements of caterpillar weight have shown that from the age IV, with the exception of the molting period, the weight of each caterpillar increases from 0.2 to 6.0 g. Already from the end of age IV, a difference in the weight of experimental (*b*) and control (*a*) caterpillars can be detected (**Figure 3**), with the former starting

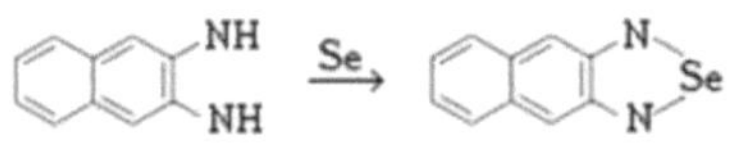

Figure 2.
In dilute solutions with 2,3-diaminonaphthalene, selenium forms fluorescent complexes, diazoselenols, with a wide spectral range, when excited by ultraviolet light.

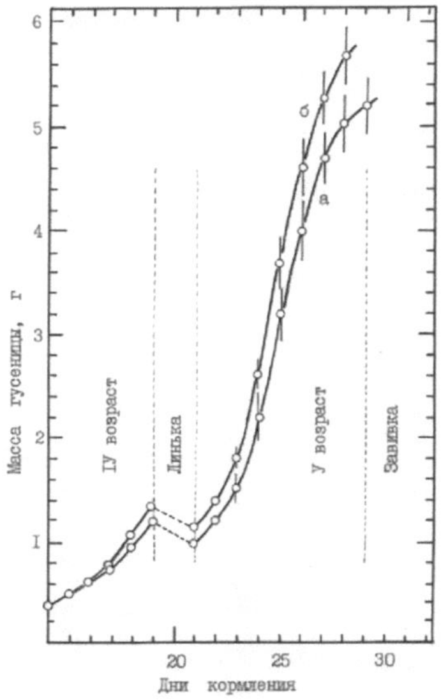

Figure 3.
Changes in mulberry silkworm caterpillar weight as a function of feeding time: A - for control batches; b - for test batches.

to curl one day earlier. This indicates that feeding the caterpillars with sodium selenite increases cellular metabolism and accelerates growth and development [104].

The effect of selenium on the growth, development, and productivity of mulberry silkworm has been studied. It is established that the yield of raw silk in experimental cocoons is 2.0–2.5% higher than in control cocoons, the metric number of yarn is better. Thus, to increase cocoon yield and improve the quality of raw silk one may recommend feeding silkworm caterpillars with sodium selenite every 48 hours at the rate of 4 micrograms of sodium selenite per gram of live weight of caterpillars from the 4th instar.

Figure 4 shows the change in selenium content in fibroin depending on the dose of sodium selenite sprayed on mulberry leaves during caterpillar feeding. The figure shows that when the dose of sodium selenite in the feed is increased to 50 μg per caterpillar, the selenium content increases from 0.04 to 0.27 mg per 1 kg of fibroin. Further increases in feed dose do not change the amount of selenium in fibroin. Consequently, *Se* has a negligible enrichment in fibroin. This indicates that not all the selenium from the feed is transferred to fibroin.

When the single dose of sodium selenite is increased above 4 mg per 1 kg live weight, caterpillar poisoning has been observed.

2.2.4 Effect of selenium on some fibroin properties

We found that when selenium is introduced into the structure of fibroin, it either replaces sulfur in the bridges between the subunits of macromolecules or forms additional lateral branching, which leads to a decrease in the rate constant of free radical formation in the matrix under the influence of UV-irradiation. In this case selenium atoms, replacing sulfur in macromolecules or forming additional branching like sulfur, lead to the capture of a great number of migrating electrons, thus reducing the rate of registered free radicals. This seems to explain the resistance of silk to radiation damage [105].

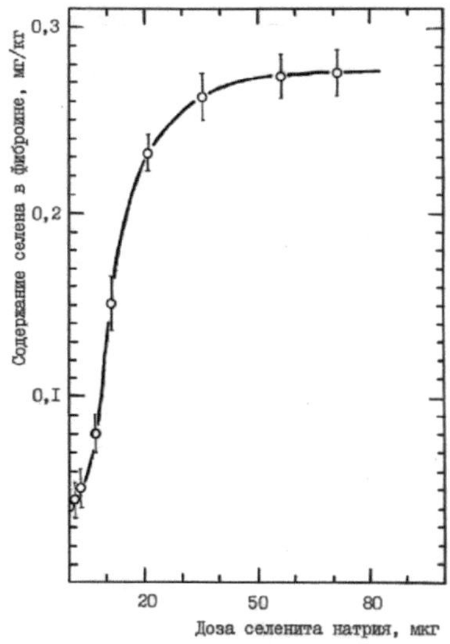

Figure 4.
Dependence of selenium content in fibroin on the dose of sodium selenite received by the silkworm caterpillar in the feed.

We investigated the effect of selenium on the time and temperature dependence of the strength of a cocoon yarn. It was found that at a constant tensile stress applied to the yarn, the value of the breaking time for the control samples was significantly lower, i.e. the strength of the control samples at a constant mechanical stress was lower than for the experimental samples. Similarly, with the same tensile time for the control specimens, the mechanical stress value is significantly higher, i.e. the control specimens withstand a higher load at a given temperature.

On the basis of the literature (S.B. Ratner, 1990) and the above experimental data on the study of the time and temperature dependence of the cocoon thread strength, as well as the nature of the material studied, it can be concluded that *Se* entering the fibroin structure changes its molecular and supramolecular structure. This, in turn, leads to a more uniform distribution of mechanical stress along the macromolecular chains, which is reflected in a reduction of the structure-sensitive parameter γ. Ultimately, the strength properties of the cocoon yarn are improved [106].

It is known that branching creates an obstacle for the proper stacking of macromolecules during their crystallization. Therefore, a change in the macro-molecular structure of fibroin when selenium is introduced should also be reflected in its supramolecular structure. Our data show that selenium introduction into fibroin structure decreases the degree of its crystallinity. This can be explained by the fact that *Se* getting into the fibroin structure forms additional branching of fibroin macromolecules. As a result, the mobility of branched macromolecules and their segments decreases during formation of the crystalline phase. Due to this slowing down, there is not enough time for the folding of the branched macromolecules and the amorphous part of the fibroin microfibrils increases [107].

To determine the nature of the change in fibroin structure following the introduction of selenium, we investigated the thermomechanical [108], deformation characteristics of fibroin [109]. In order to adequately determine the dependence of the number of amorphous sites on the concentration of selenium introduced into the fibroin, we used spin probe method, infrared spectroscopy, X-ray structure and derivatogravimetric analysis. The results are well explained by assuming that the mechanical stresses are unevenly distributed along the macromolecule chains. Selenium atoms, playing the role of a prophylactic antioxidant in fibroin, increase the resistance of the material to the effects of spark discharge. The study of these characteristics of fibroin provides qualitative information about the action of selenium, i.e. it is only indirectly possible to trace changes in the state of the amorphous sites.

It was found for the first time that during twisting of mulberry silkworm cocoon under the influence of jet stretching, caterpillar pressure, peculiarities of silk-screen structure and speed gradient crystallization of fibroin (orientation process) accompanied by formation of two modifications - CEC (crystals with elongated chains) and CFC (crystals with folded chains) occurs. Upon increasing the temperature in the derivatogravimetric chamber, crystallites with elongated fibroin chains begin to break up first, followed by crystallites with folded chains. The depth and width of DTGA minimum in low temperature region corresponding to the destruction (disordering) of EWC is much larger than EWC minimum in high temperature region. In the case of selenium-enriched fibroin, the minimum corresponding to EWC almost disappears. Thus, the introduction of selenium into the fibroin structure decreases the number of SSCs and leads to a preferential increase in the amorphous part of the polymer [110]. Fibroin is known to consist of hydrophobic and hydrophilic amino acid residues and is highly hygroscopic. It therefore quickly absorbs moisture available in the atmosphere and an equilibrium between air humidity and fibroin is established. Moisture ingress into fibroin quickly changes its electrical resistivity ϱ, polarization ε and dielectric constant $tg\delta$, which makes it possible to determine air humidity by measuring R, C and $tg\delta$. Based on these properties of fibroin, we created and patented a humidity sensor based on the selenium-enriched crystalline part of fibroin, which has a fast response and high sensitivity (M.Y. Bakirov et al. [111]). Due to the selenium content, this sensor is more resistant to aggressive environments than other materials and has a low temperature coefficient.

Author details

Shukurlu Yusif Hajibala[1*] and Huseynov Tokay Maharram[2]

1 Sheki Regional Scientific Center, National Academy of Sciences, Sheki City, Republic of Azerbaijan

2 Ministry of Science and Education, Institute of Biophysics, Republic of Azerbaijan

*Address all correspondence to: yusifsh@hotmail.com;
shrem@scence.ax

References

[1] Rayman MP. Selenium and human health. Lancet (London, England). 2012;**379**(9822):1256-1268

[2] Arthur JR, McKenzie RC, Beckett GJ. Selenium in the immune system. The Journal of Nutrition. 2003;**133**(5):1457S-1459S

[3] Wrobel JK, Power R, Toborek M. Biological activity of selenium: Revisited. IUBMB Life. 2016;**68**(2):97-105

[4] Levander OA. Scientific rationale for the 1989 recommended dietary allowance for selenium. Journal of the American Dietetic Association. 1991;**91**(12):1572-1576

[5] Schomburg L. Dietary selenium and human health. Nutrients. 2017;**9**(1):22

[6] Wisniak J, Berzelius JJ. A guide to the perplexed chemist. Chemistry and History. 2000;**5**(6):343-350

[7] Winkel LHE, Johnson CA, Lenz M, Grundl T, Leupin OX, Amini M, et al. Environmental selenium research: From microscopic processes to global understanding. Environmental Science & Technology. 2012;**46**(2):571-579

[8] Schwarz K, Foltz CM. Selenium as an integral part of faktor 3 agenist dietary necrotik liver degeneration. Journal of the American Chemical Society. 1957;**79**(12):3292-3293

[9] Hamilton JW, Tappel AL. Lipid antioxidant activity in tissues and proteins of selenium-fed animals. The Journal of Nutrition. 1963;**79**(4, 4):493-502

[10] Fairweather-Tait SJ, Bao Y, Broadley MR, Collings R, Ford D, Hesketh JE, et al. Selenium in human health and disease. Antioxidants & redox signaling. 2011;**14**(7):1337-1383

[11] Mills KW. Section-cutting of in-decalcified bone. Australian and New Zealand Journal of Surgery. 1957;**26**(4):262-266

[12] Flohe L, Gunzler WA, Schock HH. Glutathione peroxidase: A selenoenzyme. FEBS Letters. 1973;**32**(1):132-134

[13] Rotruck JT, Pope AL, Ganther HE, Swanson AB, Hafeman DG, Hoekstra WG. Selenium: Biochemical role as a component of glutathione peroxidase. 1973;**179**(4073):588-590

[14] Burk RF, Gregory PE. Some characteristics of 75se-p, a selenoprotein found in rat liver and plasma, and comparison of it with selenoglutathione peroxidase. Archives of Biochemistry and Biophysics. 1982;**213**(1):73-80

[15] Tripp MJ, Whanger PD. Association of selenium with tissue membranes of ovine and rat tissues. Biological Trace Element Research. 1984;**6**(6):455, 12-462

[16] Behne D, Wolters W. Distribution of selenium and glutathione peroxidase in the rat. The Journal of Nutrition. 1983;**113**(2):456-461

[17] Huseynov TM, Velieva NR, Khaskina PN. Comparative distribution of glutathione peroxidase activities in Guinea pig and rat tissues and organs. Reports: Academy of Sciences of Azerbaijan SSR. 1985;**41**(6):42-46

[18] Zheng H, Wei J, Wang L, Wang Q, Zhao J, Chen S, et al. Effects of selenium supplementation on graves' disease: A systematic review and meta-analysist.

Evidence-Based Complementary and Alternative Medicine. 2018;**2018**:9

[19] Labunskyy VM, Hatfield DL, Gladyshev VN. Selenoproteins: Molecular pathways and physiological roles. Physiological Reviews. 2014;**94**(3):739-777

[20] Berry MJ, Banu L, Harney JW, Larsen PR. Functional characterization of the eukaryotic secis elements which direct selenocysteine insertion at uga codons. The EMBO Journal. 1993;**12**(8):3315-3322

[21] Ye R, Huang J, Wang Z, Chen Y, Dong Y. The role and mechanism of essential selenoproteins for homeostasis. Antioxidants. 2022;**11**(5):973

[22] Zachara BA. Chapter 6.Selenium (se), blood glutathione peroxidases and DNA damage in chronic kidney disease patients on hemodialysis and after kidney transplantation - the effect of se supplementation. In: Penido MG, editor. Hemodialysis - Different Aspects. London, UK, Rijeka: IntechOpen; 2011

[23] Zachara BA, Gromadzinska J, Palus J, Zbrog Z, Swiech R, Twardowska E, et al. The effect of selenium supplementation in the prevention of dna damage in white blood cells of hemodialyzed patients: A pilot study. Biological Trace Element Research. 2011;**142**(3):274-283

[24] Behne D, Kyriakopoulos A. Mammalian selenium-containing proteins. Annual Review of Nutrition. 2001;**21**(1):453-473

[25] Saeed F, Nadeem M, Ahmed RS, Tahir NM, Arshad MS, Ullah A. Studying the impact of nutritional immunology underlying the modulation of immune responses by nutritional compounds – A review. Food and Agricultural Immunology. 2016;**27**(2):205-229

[26] Johansson L, Gafvelin G, Arn'er E.S.J. Selenocysteine in proteins—Properties and biotechnological use. Biochimica et Biophysica Acta (BBA) - General Subjects. 2005;**1726**(1):1-13

[27] Moghadaszadeh B, Beggs AH. Selenoproteins and their impact on human health through diverse physiological pathways. Physiology. 2006;**21**(5):307-315

[28] Wessjohann LA, Schneider A, Abbas M, Brandt W. Selenium in chemistry and biochemistry in comparison to sulfur. Biological Chemistry. 2007;**388**(10):997, 10-1006

[29] Arner ESJ. Selenoproteins—What unique properties can arise with selenocysteine in place of cysteine? Experimental Cell Research. 2010;**316**(8):1296-1303

[30] Kohrle J. The trace element selenium and the thyroid gland. Biochimie. 1999;**81**(5):527-533

[31] Gustin K, Vahter M, Barman M, Jacobsson B, Skroder H, Nystrom HF, et al. Assessment of joint impact of iodine, selenium, and zinc status on women's third-trimester plasma thyroid hormone concentrations. The Journal of Nutrition. 2022;**152**(7):1737-1746, 4

[32] Mayunga KC, Lim-A-Po M, Lubberts J, Stoutjesdijk E, Touw DJ, Muskiet FAJ, et al. Pregnant dutch women have inadequate iodine status and selenium intake. Nutrients. 2022;**14**(19):3936

[33] Vanderpas JB, Contempre B, Duale NL, Goossens W, Bebe N, Thorpe R, et al. Iodine and selenium deficiency associated with cretinism in northern Zaire. The American Journal of Clinical Nutrition. 1990;**52**(6): 1087-1093

[34] Demircan K, Bengtsson Y, Sun Q, Brange A, Vallon-Christersson J, Rijntjes E, et al. Serum selenium, selenoprotein p and glutathione peroxidase 3 as predictors of mortality and recurrence following breast cancer diagnosis: A multicentre cohort study. Redox Biology. 2021;**47**:102145

[35] Combs F, Jr G. Biomarkers of selenium status. Nutrients. 2015;**7**(4):2209-2236

[36] Reeves MA, Hoffmann PR. The human selenoproteome: Recent insights into functions and regulation. Cellular and Molecular Life Sciences: CMLS. 2009;**66**(15):2457-2478

[37] Kryukov GV, Castellano S, Novoselov SV, Lobanov AV, Zehtab O, Guigo R, et al. Characterization of mammalian selenoproteomes. Science. 2003;**300**(5624):1439-1443

[38] Kryukov GV, Kryukov VM, Gladyshev V. New mammalian selenocysteine-containing proteins identified with an algorithm that searches for selenocysteine insertion sequence elements. Journal of Biological Chemistry. 1999;**274**(48):33888-33897

[39] Castellano S, Morozova N, Morey M, Berry MJ, Serras F, Corominas M, et al. In silico identification of novel selenoproteins in the drosophila melanogaster genome. EMBO Reports. 2001;**2**(8):697-702

[40] Varlamova EG, Novoselov SV. Imethods to biosynthesize mammalian selenocysteine-containing proteins in vitro. Molecular Biology. 2016;**50**:697-702

[41] Bermano G, Meplan C, Mercer DK, Hesketh JE. Selenium and viral infection: Are there lessons for covid-19? British Journal of Nutrition. 2021;**125**(6):618-627

[42] Kuhn-Heid ECD, Kuhn EC, Ney J, Wendt S, Seelig J, Schwiebert C, et al. Selenium-binding protein 1 indicates myocardial stress and risk for adverse outcome in cardiac surgery. Nutrients. 2019;**11**(9):E2005

[43] Huseynov TM, Yahyayeva FR. Selenium as an Anti-Oxidative Tread in Red Blood Cells. Russian ed. London: LAP LAMBERT Academic Publishing; 2014. p. 144

[44] Huseynov TM, Yahyayeva FR, Bagirova ES, Guliyeva RT. Selenium and the problem of female reproductive health. Micronutrients in medicine. 2012;**13**(1):25-28

[45] Bonaventura C, Henkens R, Alayash AI, Banerjee S, Crumbliss AL. Molecular controls of the oxygenation and redox reactions of hemoglobin. Antioxidants & Redox Signaling. 2013;**18**(17):2298-2313

[46] Huseynova SY, Gulieva RT, Dadashov MZ, Dzhafarov AI, Yakhyaeva FR, Huseynov TM. Oxidative modification of hemoglobin of isolated erythrocytes in incubation medium containing sodium nitrite and sodium selenite (in Russian). Novosibirsk State Pedagogical University Bulletin. 2016;**6**(5):207-217

[47] Huseynova SY, Huseynov TM, Gulieva RT, Yakhyaeva FR, Dadashov MZ, Jafarov AI. Regulation of selenium oxidative processes in the blood of rats induced by sodium nitrite. Microelements in medicine. 2017;**18**(4):13-17

[48] Huseynov TM, Huseynova SY, Guliyeva RT, Dadashov MZ, Rahmanova S, Yakhyayeva FR, et al. Characteristics of oxidative stress induced by moderate doses of sodium nitrite in isolated erythrocytes in the presence of sodium selenite. Microelements in medicine (Moscow, in Russian). 2021;**22**(2):22-25

[49] Erol SA, Polat N, Akdas S, Aribal AP, Anuk AT, Ozden TE, et al. Maternal selenium status plays a crucial role on clinical outcomes of pregnant women with covid-19 infection. Journal of Medical Virology. 2021;**93**(9):5438-5445

[50] Hofstee P, Bartho LA, McKeating DR, Radenkovic F, McEnroe G, Fisher JJ, et al. Maternal selenium deficiency during pregnancy in mice increases thyroid hormone concentrations, alters placental function and reduces fetal growth. The Journal of Physiology. 2019;**597**(23):5597-5617

[51] Lewandowska MS, Sajdak S, Lubin'ski J. Serum selenium level in early healthy pregnancy as a risk marker of pregnancy induced hypertension. Nutrients. 2019;**11**(5):1028-1042

[52] Chengmin W, Haijing W, Jing L, Yi H, Lei W, Mingxing D, et al. Selenium deficiency impairs host innate immune response and induces susceptibility to listeria monocytogenes infection. BMC Immunology. 2009;**10**(1):52

[53] Kehinde SO, Olusola FO, Gbemisola EO, Sarah JO, Sulaimon AA, Rose IA. Selenium deficiency and pregnancy outcome in pregnant women with hiv in Lagos, Nigeria. International Journal of Gynaecology and Obstetrics. 2018;**142**(2):207, 8-213

[54] Gusev VA. Free-radical theory of aging in the paradigm of gerontology (in russian). Medline.ru. Medical and Biological Sciences: Advances in Gerontology. 2002;**3**:271-272

[55] Harman D. Aging: A theory based on free radical and radiation chemistry. Journal of Gerontology. 1956;**11**(3):298-300

[56] Savarino L, Granchi D, Ciapetti G, Cenni E, Ravaglia G, Forti P, et al. Serum concentrations of zinc and selenium in elderly people: Results in healthy nonagenarians/centenarians. Experimental Gerontology. 2001;**36**(2):327-339

[57] Steinbrenner H, Sies H. Protection against reactive oxygen species by selenoproteins. Biochimica et Biophysica Acta (BBA) - General Subjects. 2009;**1790**(11):1478-1485 Special Issue: Selenoprotein Expression and Function

[58] Brigelius-Flohe R, Matilde M. Glutathione peroxidases. Biochimica et Biophysica Acta. 2013;**1830**(5):3289-3303

[59] Viktorovich GA, Lyapunov SM, Okina OI, Frontasyeva MV. The intake of selenium and iodine in the human body with different diets (in Russian). Food and Nutrition Sciences. 2011;**10**:3-8

[60] Wong CKH, Lang BHH, Lam CLK. A systematic review of quality of thyroid-specific health-related quality-of-life instruments recommends thypro for patients with benign thyroid diseases. Journal of Clinical Epidemiology. 2016;**78**:63-72

[61] Ventura M, Melo M, Carrilho F. Selenium and thyroid disease: From pathophysiology to treatment. International Journal of Endocrinology. 2017;**1**:31

[62] Huseynova SY, Gulieva RT, Yahyaeva FR, Huseynov TM. Selenium and thyroid disease: From pathophysiology to treatment. Biomedicine (Baku). 2019;**17**(2):4-12

[63] Schomburg L. Selenium, selenoproteins and the thyroid gland: Interactions in health and disease. Nature Reviews Endocrinology. 2012;**8**:160-171

[64] Wichman J, Winther KH, Bonnema SJ, Hegedu¨s L. Selenium

supplementation significantly reduces thyroid autoantibody levels in patients with chronic autoimmune thyroiditis: A systematic review and meta analysis. Thyroid. 2016;**26**(12):1681-1692

[65] Leopold F. Selenium and Human Health: Snapshots from the Frontiers of Selenium Biomedicine. Berlin Heidelberg: Springer; 2011. pp. 285-302

[66] Duntas LH, Salvatore B. Selenium: an element for life. Endocrine. 2015;**48**:756-775

[67] Romanenko TG, Chaika OI. Peculiarities of thyroid function in pregnant women with iodine deficiency. International Journal of Endocrinology. 2014;**60**:89-94

[68] Huang Z, Rose AH, Hoffmann PR. The role of selenium in inflammation and immunity: From molecular mechanisms to therapeutic opportunities. Antioxidants & Redox Signaling. 2012;**16**(7):705-743

[69] Avery JC, Hoffmann PR. Selenium, selenoproteins, and immunity. Nutrients. 2018;**10**(9):1203

[70] Hoffmann FW, Hashimoto AC, Shafer LA, Dow S, Berry MJ, Hoffmann PR. Dietary selenium modulates activation and differentiation of CD4+ T cells in mice through a mechanism involving cellular free thiols. The Journal of Nutrition. 2010;**140**(6):1155-1161

[71] Sneha H, Selvakumar D. Selenium and selenoproteins: Its role in regulation ofinflammation.Inflammopharmacology. 2020;**28**:667-695

[72] Vinceti M, Filippini T, Del Giovane C, Dennert G, Zwahlen M, Brinkman M, et al. Selenium for preventing cancer. Cochrane Database of Systematic Reviews. 2018;**1**(1):CD005195

[73] Shamberger RJ, Willis CE. Selenium distribution and human cancer mortality. CRC Critical Reviews in Clinical Laboratory Sciences. 1971;**2**(2):211-221

[74] Jafarov AI, Huseynov TM, Jamalov DR, Mehtiyev MA, Teplyakova GV. Changes of Selenium Content in Rat Tissues during Development of Malignant Tumors. Baku: Elm; 1974. pp. 129-131

[75] DeMartino AW, Kim-Shapiro DB, Patel Rakesh P, Gladwin MT. Nitrite and nitrate chemical biology and signalling. British Journal of Pharmacology. 2019;**176**(2):228-245

[76] Sies H, Klotz LO, Sharov VS, Assmann A, Briviba K. Protection against peroxynitrite by selenoproteins. Zeitschrift fur Naturforschung. C. Journal of Biosciences. 1998;**53**(3-4):228-232

[77] Storkey C, Pattison DI, Ignasiak MT, Schiesser CH, Davies MJ. Kinetics of reaction of peroxynitrite with seleniumand sulfur-containing compounds: Absolute rate constants and assessment of biological significance. Free Radical Biology & Medicine. 2015;**89**:1049-1056

[78] Sies H, Arteel GE. Interaction of peroxynitrite with selenoproteins and glutathione peroxidase mimics. Free Radical Biology & Medicine. 2000;**28**(10):1451-1455

[79] Suzuki KT, Shiobara Y, Itoh M, Ohmichi M. Selective uptake of selenite by red blood cells. The Analyst. 1998;**123**(1):63-67

[80] Haratake M, Fujimoto K, Hirakawa R, Ono M, Nakayama M. Hemoglobin-mediated selenium export from red blood cells. JBIC Journal of Biological Inorganic Chemistry. 2008;**13**(3):471-479

[81] Beilstein MA, Whanger PD. Distribution of selenium and glutathione peroxidase in blood fractions from humans, rhesus and squirrel monkeys, rats and sheep. The Journal of Nutrition. 1983;**113**(11):2138-2146

[82] Singel DJ, Stamler JS. Chemical physiology of blood flow regulation by red blood cells: The role of nitric oxide and snitrosohemoglobin. Annual Review of Physiology. 2005;**67**(1):99-145

[83] Vitturi DA, Sun C-W, Harper VM, Thrash-Williams B, CantuMedellin N, Chacko BK, et al. Antioxidant functions for the hemoglobin 93 cysteine residue in erythrocytes and in the vascular compartment in vivo. Free Radical Biology and Medicine. 2013;**55**:119-129

[84] Mansouri A. Oxidation of human hemoglobin by sodium nitrite-effect of-93 thiol groups. Biochemical and Biophysical Research Communications. 1979;**89**(2):441-447

[85] Huseynov TM, Safarov NS. Selenium and certain viral diseases. Journal of Biomedicine (Journal of Azerbaijan Medical University). 2007;**2**:3-4

[86] Huseynov TM, Guliyeva RT, Jafarova SH, Jafar NH. Sodum selenite as a potential adjuvant therary for covid-19. Journal of Biophysics (Russian Academy of Sciences). 2022;**67**(5):956-961

[87] Maltseva VN, Goltyaev MV, Turovsky EA, Varlamova EG. Immunomodulatoryandanti-inflammatory properties of selenium-containing agents: Their role in the regulation of defense mechanisms against covid-19. International Journal of Molecular Sciences. 2022;**23**(4):2360

[88] Lu R, Zhao X, Li J, Niu P, Yang B, Wu H, et al. Genomic characterisation and epidemiology of 2019 novel coronavirus: Implications for virus origins and receptor binding. Lancet (London, England). 2020;**395**:565-574

[89] Mittal A, Manjunath K, Ranjan RK, Kaushik S, Kumar S, Verma V. Covid-19 pandemic: Insights into structure, function, and hace2 receptor recognition by sars-cov-2. PLoS Pathogens. 2020;**16**(8):e1008762

[90] Diwaker D, Mishra KP, Ganju L. Potential roles of protein disulphide isomerase in viral infections. Acta Virologica. 2013;**57**(3):293-304

[91] Kieliszek M, Lipinski B. Selenium supplementation in the prevention of coronavirus infections (covid-19). Medical Hypotheses. 2020;**143**:109878

[92] Younesian O, Khodabakhshi B, Abdolahi N, Norouzi A, Behnampour N, Hosseinzadeh S, et al. Decreased serum selenium levels of covid-19 patients in comparison with healthy individuals. Biological Trace Element Research. 2022;**200**(4):1562-1567

[93] Dailey GP, Premadasa LS, Ruzicka JA, Taylor EW. Inhibition of selenoprotein synthesis by zika virus may contribute to congenital zika syndrome and microcephaly by mimicking selenop knockout and the genetic disease pcca. BBA Advances. 2021;**1**:100023

[94] Taylor EW, Ruzicka JA. Antisense inhibition of selenoprotein synthesis by zika virus may contribute to neurological disorders and microcephaly by mimicking sepp1 knockout and the genetic disease progressive cerebello-cerebral atrophy. Bulletin of the World Health Organization. 2016;**10**:7

[95] Suzuki KT. Metabolomics of selenium: Se metabolites based on speciation studies. Journal of Health Science. 2005;**51**(2):107-114

[96] Sumida M, Sutthikhum V. Fibroin and sericin-derived bioactive peptides and hydrolysates as alternative sources of food additive for promotion of human health: A review. Research Knowledge. 2015;**1**(2):1-17

[97] Minghong L, Shenzhou L, Xilong W. Effect of glycerol's water solubility and structure on silk fibroin films (patent: Cn102516777b). Chemical Journal of Chinese Universities. 2011;**11**:29

[98] Wen C, He X, Zhang J, Liu G, Xu X. A review on seleniumenriched proteins: Preparation, purification, identification, bioavailability, bioactivities and application. Food and Function. 2022;**13**:5498-5514

[99] Puccinelli M, Malorgio F, Pezzarossa B. Selenium enrichment of horticultural crops. Molecules. 2017;**22**(6):933

[100] Abdullayev HMB, Bakirov MY, Alizade M, Khalilov M, Abdurakhmanova MD, Efendiev ST. Author's certificate no. 481279, (ussr) Growth Stimulant for Mulberry Silkworms. Vol. 121975. Available from: https://patenton.ru/patent/SU481279A1

[101] Yamada H, Nakao H, Takasu Y, Tsubouchi K. Preparation of undegraded native molecular fibroin solution from silkworm cocoons. Materials Science and Engineering: C. 2001;**14**(1):41-46

[102] Lindstr¨om K. Selenium as a Growth Factor for Plankton Algae in Laboratory Experiments and in some Swedish Lakes. Netherlands, Dordrecht: Springer; 1983. pp. 35-47

[103] Walkinson JH. Fluorometric determination of selenium in biological material with 2,3-diaminonaphthalene. Analytical Chemistry. 1966;**38**:92-97

[104] Bakirov MY, Shukurlu YH. Silkworm growth stimulator. "Shelk" (Scientific and Technical Abstracts). 1981;**2**(95):21-22

[105] Abdullayev HB, Bakirov MY, Mammadov SV, Khalilov ZM, Shukurlu YH, Yusifov EY. Study of the effect of selenium on the kinetics of free radical accumulation in fibroin. Reports: Academy of Sciences of Azerbaijan SSR. 1978;**34**(11):20-24

[106] Abdullayev HB, Bakirov MY, Mammadov SM, Abasov SA, Shukurlu YH. Study of the effect of selenium on the time and temperature dependence of cocoon yarn strength. Proceedings of the Academy of Sciences of the Azerbaijan SSR : Series of Physical and Technical Sciences and Oil. 1980;**2**:86-89

[107] Shukurlu YH, Kerimov TM, Bakirov MY, Mammadov SV. Investigation of selenium effect upon the ultramolecular structure of the natural silk. Proceedings of the Academy of Sciences of the Azerbaijan SSR: Series of Physical and Technical Sciences and Oil. 1981;**1**:110-113

[108] Shukurlu YH, Bakirov MY, Mammadov SV. Installation for determination of thermomechanical curves of natural silk. "Shelk" (Scientific and Technical Abstracts). 1980;**6**(93):21-22

[109] Bakirov MY, Shukurlu YH, Ismailova RS, Mammadov SV. Combined effect of uv irradiation and selenium on deformation and strength properties of natural silk. "Shelk". Scientific and Technical Abstracts. 1981;**4**(97):27-28

[110] Shukurlu YH. The effect of selenium on the supramolecular structure and thermal characteristics of fibroin bombyx mori l. Periodico Tche Quimica (Online). 2020;**17**(552119593):591, 12-598

[111] Bakirov MY, Shukurlu YH, Ismailova RS, Mammadov SV. Humidity Sensor (Patent Ussr). Vol. 71986

Section 3

Effects of Selenium and Its Components on Human Health

Chapter 8

Plant-Based Foods Biofortified with Selenium and Their Potential Benefits for Human Health

Soledad García-Morales, Janet María León-Morales, Víctor García-Gaytán and Luis Guillermo Terreros-Rosales

Abstract

Selenium (Se) is essential for humans. This element is present in more than 25 proteins related to redox processes, and its deficiency is related to the onset of chronic diseases. One way to incorporate Se into the human diet is by consuming plant foods rich in Se. Crop fortification with Se can be achieved through the agronomic practice of biofortification. This chapter discusses dietary sources of inorganic Se (selenate and selenite), organic Se (selenocysteine, selenomethionine, and methylselenocysteine), and bioactive compounds provided by consuming the edible parts of plants as a result of agronomic biofortification. The benefits to human health from consuming selenium-enriched crops due to their biological functions such as antioxidant, anti-inflammatory, and anticarcinogenic are also presented. The intake of Se-enriched plant foods is a growing trend. In addition to providing the daily dose of Se, these Se-enriched vegetables are a functional food option that improves human health due to their content of phytochemical compounds.

Keywords: biofortification, inorganic Se, organic Se, bioactive compounds, antioxidant, anti-inflammatory, anticarcinogenic

1. Introduction

Selenium (Se) is an element that is required in trace amounts and has an essential role in human metabolism, growth, and hormonal balance [1]. In humans, 25 selenoproteins have been reported and classified into six functional groups (proteins involved in Se transport, selenocysteine synthesis, protein folding, hormone metabolism, redox signaling, and reductase/peroxidase activity) [2]. Although most selenoproteins are related to protection against oxidative stress, others are involved in phospholipid biosynthesis and calcium signaling [2]. Selenium has also been reported to intervene in health through epigenetic processes, modulating DNA methylation and histone acetylation [3].

Meanwhile, Se deficiency can lead to human health problems ranging from endemic cardiomyopathy (Keshan disease), endemic deforming osteoarthropathy (Kashin-Beck disease), male infertility, prostate cancer, cystic fibrosis, muscular dystrophy and impairment of the immune system, and reducing defenses against infectious viral diseases (influenza, hepatitis, HIV or SARS-CoV-2) [4].

Selenium is characterized by its ability to transition to different oxidation states. In nature, Se has five oxidation states (+6, +4, 0, −1, and − 2) and different selenate (Se^{+6}, SeO_4^{2-}, Se (VI)), selenite (Se^{+4}, SeO_3^{2-}, Se(IV)), elemental Se (Se^0), and selenide (Se^{2-}) forms, in addition to its organic forms such as selenocysteine (SeCys), selenomethionine (SeMet), and methylselenocysteine (MeSeCys) [3, 4]. These forms of Se are commonly found in traditional dietary supplements, along with selenized yeast rich in SeMet. Meanwhile, Se in proteins is found in the form of the amino acids SeCys and SeMet [3].

Although Se is an essential element for humans, its biological activity and bioavailability depend on a number of factors such as chemical form, accessibility, solubility, digestibility, the amount ingested, and physiological state of the organism, as well as the presence of other components in the diet [3]. Studies have revealed that the organic forms of Se are less toxic and are absorbed more efficiently than the inorganic forms of Se. Of the latter, Se^{+4} is more toxic than Se^{+6} [3]. In turn, Niedzielski et al. [5] indicate that organic Se compounds have a higher bioavailability and are assimilated in ranges of 85–95% when it comes to food/supplements, whereas inorganic selenium has an absorption range of 40–50% during human intake.

The recommended dietary allowance of Se for humans depends on gender, age, pregnancy, lactation, dietary intake, and geographical location. The United States (US) Department of Agriculture indicates a dose of 55 μg/day as the recommended daily allowance (RDA), while the European Food Safety Authority (EFSA) indicates an RDA of 70 μg/day for men, 60 μg/day for women, and 75 μg/day for lactating women, being a more specific dose. Meanwhile, the US Institute of Medicine expert panel determined the tolerable upper limit (UL) at 400 μg/day and the no-observed-adverse-effect level (NOAEL) at 800 μg/day [3]. Finally, the International Food and Nutrition Board suggests an average daily intake of 40–70 μg/day for men, 45–55 μg/day for women, and 25 μg/day for children [3]. Therefore, it is important to maintain a balance in the daily dose of Se, since doses higher than 1.2 mg/day can cause toxic effects and lead to neurophysiological alterations (confusion, memory loss, dizziness, irritability, fatigue, anxiety, anger, insomnia, depression, or headache), eye problems, skin lesions, or hair and nail loss [3, 4].

The production of Se-enriched plant foods can be an alternative to the consumption of biofortified vegetables to reduce Se deficiency, thus preventing and treating several diseases that threaten human health [6]. In recent years, Se biofortification has emerged as an effective strategy to increase the Se content in crops and thus improve its availability in the edible parts of cultivated plants, allowing this trace element to enter the food chain and strengthen human health.

2. Agronomic biofortification

For the agronomic biofortification of Se in plants, research has been generated in terms of concentration, type of plants, dynamics and different forms of Se in the soil, type of crops, application methods, and lately its nanotechnological use in agriculture. Se biofortification consists of a process to increase the bioavailability of Se, in plants consumed during human intake, without compromising crop yields [7]. This strategy can be achieved by agronomic techniques or through gene targeting [8]. The main agronomic methods for Se biofortification are foliar applications and soil applications, with foliar spraying of Se being the most efficient because this prevents selenate leaching and selenite fixation in the soil [9]. Selenite and selenate are the two inorganic forms of Se that are mainly used as fertilizers for the exogenous application of Se to plants. Currently, there are other agronomic techniques for Se biofortification such as Se-enriched nutrient solution in hydroponics and seed soaking, among others [8, 10].

2.1 Selenium biofortification in hydroponic systems

The technological approach of soilless cultivation seems to be a key factor for strict control of crop conditions and observation of the effect of Se in a biofortification strategy. Through this system, with the joint addition of Selenium (Se) + Iodine (I), there was an activation of the biosynthesis of organic forms of Se. In leaf vegetables such as lettuce, it was shown that the application of Se + I, with a low dose of salicylic acid, increased the sugar content in leaves and improved the concentration of macro- and micronutrients in roots (P and Mn) [11, 12]. The addition of 5 μM Se to the nutrient solution could be considered a high concentration but at the same time safe for human and plant health as it stimulated lettuce growth and yield and increased the content of phenolic compounds [13]. Under hydroponic conditions, supplying Se to the nutrient solution delayed and reduced the toxic effects of cadmium (Cd) on bell pepper plants [14]. In another study, humic/fulvic acid mixture plus root application of Se in the nutrient solution reduced the harmful effects caused by Cd toxicity in broccoli plants; furthermore, improvements in growth rate and reduction in Cd transport from leaves to inflorescence were observed [15]. Selenium appears to positively affect cell membrane stability in cucumber plants exposed to Cd, as Cd accumulation in roots was reduced [15]. In addition, selenoproteins act as antioxidant agents in plant metabolism, increasing the activity of enzymatic and non-enzymatic compounds that together act against reactive oxygen species (ROS) and cellular detoxification [16].

2.2 Selenium biofortification in soil and foliar spray crops

The joint foliar application of Se + I is an interesting biofortification method, although this strategy presents some difficulties due to the toxicity of Se [17]. Although it is a very efficient method for product application, it was observed that foliar application of Se did not reduce the toxic effects of Cd on bell pepper plants; whereas, root application with nutrient solution proved to be a more effective method [14]. It is not recommended to apply Se to broccoli plants to mitigate the toxic effects of Cd, as this could further increase its toxicity [15]. Foliar application of a micronutrient mixture (zinc (Zn), iron (Fe) I, and Se) represented an effective strategy for wheat biofortification, without yield effects [18]. This micronutrient mixture also had beneficial effects on rice grain, as the Zn, I, and Se content was increased [19]. A high dose of Se (10 mg/kg) decreased grain yield and biomass in wheat. Whereas, Se (in the form of selenite) accumulated mainly in wheat grain and root, a higher accumulation in the form of selenate was found in leaf and straw [20]. Selenium is chemically similar to sulfur (S) and is taken up by plants through S transporters present in the root plasma membrane, metabolized by the S assimilation pathway, and volatilized to the atmosphere [21]. Plants can take up inorganic Se (selenate, selenite, or elemental Se) and organic Se (SeCys and SeMet); the forms and availability of Se will depend on soil type and pH [22]. For biofortification, it is necessary to consider many factors, the method of application, the timing of application, the pH of the mixture, and the concentrations, and to know the possible synergistic and antagonistic effects between the products to be applied [23].

It is also worth mentioning that there are new nanotechnological tools for agronomic biofortification. A study revealed that Se nanoparticles (SeNPs) could be used for Se supplementation, an essential microelement for humans. With the application of 4.65 μg/mL SeNPs, the highest germination percentage was obtained in barley seeds [24]. A field experiment revealed that SeNPs improved growth parameters, carotenoid content, and insect control in sunflowers when 20 mg/L was applied [25]. SeNPs increased the activity of enzymes related to free radical scavenging; in addition, SeNPs showed excellent bioavailability, low toxicity, and high biological activity [26]. In tomato, Se application significantly favored the tomato fruit quality, including total soluble solids, soluble sugar, and titratable acid [27]. The use of Se-pelleted seeds has emerged as an interesting and viable alternative to increase Se supplementation in agricultural ecosystems [28].

3. Source of inorganic and organic selenium from the crop plant

Because organisms cannot synthesize Se, humans enter Se into their diet mainly through the intake of cultivated foods, so one strategy to increase Se content in crops and the human food chain is through agronomic biofortification with Se. It is also important to understand the bioaccessibility of Se in the edible tissues of Se-enriched crops.

Selenoproteins are the form in which Se is present in the human body; for this purpose, Se can be ingested in organic and inorganic forms. Selenite and selenate (organic Se) and methionine (organic Se) are considered highly bioavailable. Elemental Se is classified as difficult to be absorbed by the gastrointestinal tract. In addition, organic Se from food intake is considered relatively safe for the human body, whereas inorganic Se ingested by chemical supplements has a narrow range between its therapeutic effect and its toxic potential [1].

One of the crops that stands out for its consumption throughout the world is wheat, which is also characterized by its ability to accumulate Se. In a study by Wang et al. [10], it was found that regardless of the method of biofortification (foliar or soil application) and the form of exogenous Se applied (selenite or selenate), the speciation of Se in wheat grains was the organic form (93–100%). Organic Se in wheat grains comprised 87–96% SeMet and 4–13% $SeCys_2$; whereas, the inorganic Se species was selenate (1–6%). The bioaccessibility of Se in white flour and whole wheat flour was also determined in this study. In the intestinal phase, 10–38% bioaccessibility was reported in white wheat flour and 9–34% in whole wheat flour, while in the gastric phase, Se bioaccessibility was similar between white flour (6–34%) and whole wheat flour (6–27%) [10].

Rice is considered the staple food for more than half of the world's population, making it a strategic crop for biofortification and Se intake. In an analysis of Se speciation in rice grains, where the application of selenite to the soil and by foliar spraying was evaluated, it was found that selenite was the dominant Se species (≈42–73%), the inorganic Se species being the prevailing one in rice grains and the organic species being a smaller proportion. A strong influence of the biofortification method was also reported; root application of selenite favored the presence of seleno-amino acids (≈38%), and foliar spraying induced the accumulation of selenite (≈73%) and selenate (≈15%) in rice grains [29]. Se speciation changed when dealing with brown rice grains biofortified with foliar application of selenite, where SeMet was the main metabolite identified in Se-enriched rice [30].

The third-most consumed crop in the world is potato. During the production of potato tubers, selenite or selenate was applied by foliar spraying during different stages of plant growth. The main organic Se species in potato tubers was SeMet (78.6% with selenite application and 52.3% with selenate), although the presence of SeCys2 and SeMeCys was also detected. Selenate was the predominant inorganic Se species, and its proportion varied according to Se source (1.5% with selenite and 31.9% with selenate) [9].

Se biofortification has also been studied in other cereals such as maize, in legumes such as cowpea, as well as in other crops such as groundnut. These crops had a high proportion of organic Se (>90%), indicating that the plants were highly efficient in converting inorganic Se to organic Se. SeMet was the dominant organic Se species in all three crops with proportions of 92.0% in maize, 63.7% in cowpea, and 85.2% in groundnut. $SeCys_2$ was also identified (7.1% in maize, 2.1% in cowpea, and 10.4% in groundnut). Cowpea grains stood out from the other two crops for their MeSeCys content (31.7%). As for inorganic Se species, the proportion was 2.7% selenate in cowpea and 2.1% selenite in groundnut. Gastrointestinal bioaccessibility was also determined in this work, and a range of 66.6–78.2% was found for the three crops, with no differences among the three types of grains enriched with Se [31].

In peanut, foliar and soil application (root irrigation) of selenite was evaluated, and the main Se species in peanut protein were determined. The major organic Se species was SeCys2 (65.3%), followed by MeSeCys (13.9%); the inorganic form of Se was selenite and accounted for 11.7% of the total Se compounds. The organic Se content in peanut was about 86.3%. This crop efficiently absorbed and transformed selenite into organic Se sources [32].

The ability of strawberry plants to absorb and biosynthesize inorganic Se into seleno-amino acids has also been studied, with foliar application of selenite being the best biofortification treatment compared to other Se sources such as selenate or SeMet applied in root irrigation. In strawberry fruits, 86% of the total Se content is identified, and 16% corresponds to two unknown Se species. Of the identified Se species, 45% corresponds to SeMet, 20.7% to MetSeCys, 5.8% to SeCyts, 5.6% to selenite, and 6.6% to selenate [33].

In the case of vegetables such as lettuce, four Se species were detected, SeMet, SeCys, selenite, and selenite. The proportion of these species was a function of the Se source used in biofortification. With selenate application, the proportion of SeMet, SeCys, and selenite was 51%, 4%, and 45%, respectively. Meanwhile, with selenite treatment, 90% of SeMet, 10% of SeCys, and no record of inorganic Se was obtained, indicating that all the supplemented selenite was converted into organic Se. In edible lettuce shoots, regardless of the source of Se applied, the proportion of organic Se was higher than the proportion of inorganic Se [8].

Sprouts are seedlings from seeds, which, after germination, are consumed with fresh vegetables. These types of plant foods are gaining interest because they may contain more bioactive compounds than seeds and can be enriched with Se. In the case of soybean sprouts, two Se sources (Se nanoparticles (SeNPs) and selenite) and two concentrations were evaluated. With the application of SeNPs, five Se species were identified in soybean sprouts, the organic Se species SeMet (55–71%), SeCys2 (6–17%) and MeSeCys (6–14%) as well as the inorganic Se species selenite (2%) and selenate (11.5–15%). Whereas, in selenite-enriched soybean sprouts, SeMet species predominated (71.5–89-1%), followed by $SeCys_2$ (4.5–14.4%), MeSeCys (4.2–10.4%), and selenite (2.3–3.7%) [34].

Crop	Edible plant	Selenium (Se) speciation in crops		Reference
		Inorganic Se	Organic Se	
Wheat	Grain	Selenate: 1–6%	Selenomethionine (SeMet): 87–96%	[10]
			Selenocysteine (SeCys): 4–13%	
Rice	Grain	Selenate: ≈15–18%	≈8–37%	[29]
		Selenite: ≈42–73%		
Maize	Grain		SeMet: 92%	[31]
			SeCys: 7.1%	
			Methylselenocysteine (MeSeCys): 0.9%	
Cowpea	Grain	Selenate: 2.7%	SeMet: 63.7%	
			SeCys: 2.1%	
			MeSeCys: 31.7%	
Groundnut	Grain	Selenite: 2.1%	SeMet: 85.2%	
			SeCys: 10.4%	
			MeSeCys: 2.2%	
Papa	Tuber	Selenate: 1.5–31.9%	SeMet 50–80%	[9]
Peanut	Grain	Selenite: 11.7%	SeCys: 65.3%	[32]
			MeSeCys: 13.9%	
Strawberry	Fruit	Selenite: 5.6%	SeMet: 45%	[33]
		Selenate: 6.6%	SeCys: 5.8%	
			MeSeCys: 20.7%	
Lettuce	Shoot	Selenite: 0–45%	SeMet: 51–90%	[8]
			SeCys: 4–10%	
Soybean	Sprouts	Selenite: 2.1–3.7%	SeMet: 55.1–89.1%	[34]
		Selenate: 11.5–15%	SeCys: 4.5–17.3%	
			MeSeCys: 4.2–13.9%	

Table 1.
The proportion of Se species in edible organs of different crops biofortified with Se.

Plants have the ability to uptake and metabolize Se, which makes them ideal Se sources for daily dietary Se supplementation. Many crop plants have been shown to have a high capacity to convert inorganic Se into organic Se. In plants, Se species are related to the type of crop; thus, different crops may have different inorganic or organic Se species (**Table 1**). The organic Se species are seleno-amino acids such as selenocysteine (SeCys) and selenomethionine (SeMet), which in turn can give rise to methylated SeCys (MeSeCys) and methylated SeMet (MeSeMet). These organic forms of Se have bioactive properties that benefit human health as anticarcinogens and in the regulation of inflammatory processes.

Recently, the amount of research on Se biofortification has focused on crop production; of these, cereals are the ideal crops for Se biofortification due to their high consumption worldwide. However, the cultivation of hydroponic vegetables, such as lettuce, and the production of sprouts are also excellent options because they have a short production cycle, are easy to handle, have fresh taste characteristics, and can be eaten fresh or cooked [8]. These vegetables along with fresh fruits, such as strawberries, can frequently be found in the diet of people around the world. To date, there has been a great diversity of Se-enriched plant foods that can be ingested to supplement the Se required by the human body to maintain or improve health.

4. Secondary metabolites derived from Se biofortified crops

Phytochemicals or secondary metabolites have no recognized role in the vital processes of plants but are important in their interaction with the environment. From the point of view of human health, there is extensive evidence of the diverse biological activities presented by the different classes of phytochemicals, which include antioxidant, anti-inflammatory, antimicrobial, antitumor, and immunomodulatory, among others. Therefore, in recent years, there has been growing interest in the consumption of vegetables rich in these bioactive compounds for the prevention of chronic diseases and the regulation of oxidative stress [35]. The production of these phytochemicals can be elicited in response to biotic (bacteria, fungi, viruses) and abiotic (drought, salinity, heavy metals, UV radiation) stress factors.

In several studies, it has been observed that biofortification with Se is useful to increase the content of this trace element in the edible parts of plants as well as improves their nutraceutical value through the accumulation of biocompounds. In addition to the beneficial health properties, these phytochemicals also provide fruits and inflorescences with their organoleptic properties, such as lycopene in tomato, capsaicin in chili, and glucosinolates in broccoli.

The application of Se in foliar form, as a soil amendment, in the irrigation solution, or in hydroponics has a positive effect on the accumulation of phenolic compounds, terpenes, capsaicinoids, and glucosinolates. The accumulation of phenolic compounds in response to Se has been extensively evaluated, in some plant species, by determining their total content and in others, by identifying some compounds individually, in different plant organs (**Table 1**). In bean grains, root irrigation application of 5 and 10 μM Na_2SeO_3 increased the content of total phenolic compounds and total flavonoids differentially among common bean varieties [36]. In lettuce leaves, the tentative identification and quantification of caffeoylquinic and dicaffeoyltartaric acids, as well as glycosylated derivatives of quercetin and cyanidin, was carried out. From a concentration of 0.04 mg/L Na_2SeO_4, an increase in the response of these phytochemicals was observed by electrospray ionization mass spectrometry (ESI-MS) [6]. In basil leaves, increases in the content of different phenolic acids (gallic, chlorogenic, coumaric, rosmarinic acids) were achieved with the application of 50 mg/L SeNPs, but in the case of caffeic acid, a positive response was only observed at twice the concentration [37]. The use of Se nanomaterials as a base fertilizer in soil for lettuce cultivation induced increases in the abundance of quercetin (2.9-fold), rutin (2.7-fold), and coumarin (2.4-fold) [38]. In jalapeño pepper fruits, the content of phenolic compounds and total flavonoids increases as higher Na_2SeO_3 concentration is applied and correlates with the observed antioxidant capacity [39]. Selenium, in the form of Na_2SeO_4, also stimulated the production of phenolic compounds, flavonoids, and anthocyanins, as well as the expression of biosynthetic enzymes (phenylalanine ammonium lyase and chalcone synthase) in Indian mustard leaves [40]. In microgreens biofortified with Na_2SeO_4, the most abundant phenolics are chlorogenic acid and rutin (coriander), caffeic acid hexoside and kaemferol-3-O (caffeoyl)-sophoroside 7-O-glucoside (tatsoi), and chicoric and rosmarinic acids (basil) [41].

In broccoli florets, the Se source is important in the induction of these phytochemicals, obtaining positive effects on the production of phenolic acids with the lowest doses of Se yeast, while Na_2SeO_3 had similar effects only with the highest doses (**Table 2**). In contrast, flavonoid content increased with the highest Na_2SeO_3 concentration but did not undergo any modification when the organic Se source was applied. In the case of glucosinolates, both Se sources induce their accumulation [42].

Induction of secondary metabolism by Se can be carried out by increasing the content of this element in the same vegetative organ (direct) or even in an indirect way. Se accumulation in broccoli florets as the dose of Na_2SeO_4 (applied to roots) increases causes contrasting effects on two classes of phytochemicals; at the intermediate concentration evaluated (0.4 mmol/L), Se induced glucosinolate production and reduced flavonoid content [44]. Similarly, in cauliflower, foliar application of Na_2SeO_4 results in the accumulation of this element in florets, inducing a higher content of carotenoids and phenolic compounds in two cultivars. The Graffiti cultivar accumulated twice as many glucosinolates as the Clapton cultivar at the 5 mg/L doses, identifying glucobrassicin, 4-hydroxy glucobrassicin, 4-methoxy glucobrassicin, and neo-glucobrassicin [45]. In tomato fruits, this direct induction of Se on flavonoid content is also observed, with no change in lycopene content [46]. However, with the foliar application of 1.5 mg/L Na_2SeO_3, no changes in the accumulation of this trace element in jalapeño pepper fruits were recorded, but an increase in the content of flavonoid, phenolic compounds, and capsaicin was noted [39]. Therefore, it is relevant to carry out studies on the mechanism by which this trace element induces the synthesis of these bioactive compounds.

Plants are naturally exposed to several stress factors simultaneously. In this sense, some studies have evaluated the effect of Se in combination with other elements or stressors on the accumulation of phytochemicals (**Table 2**). In tea leaves, the enzymatic (SOD) and non-enzymatic (epigallocatechin and epigallocatechin gallate) antioxidant systems are activated in response to Se, which may be part of the strategy

Species	Plant part	Biofortification method	Bioactive compounds	Reference
Phaseolus vulgaris L.	Grains	Root irrigation every 15 d: 0, 2.5, 5, and 10 μM Na_2SeO_3.	Total phenolic compounds and total flavonoids	[36]
Lactuca sativa L.	Leaves	Weekly foliar application for 3 weeks: 0, 0.04, and 0.5 mg Na_2SeO_4 /L	Phenolic acids, flavonoids, and sesquiterpene lactones	[6]
Lactuca sativa L.	Leaves	Soil amended with selenium nanomaterials: 0, 0.1, 0.5, and 1.0 mg /kg; and 0.5 mg SeO_3^{-2}/kg	Quercetin, rutin, caffeic acid, and coumarins	[38]
Lactuca sativa L.	Leaves	Nutrient solution: 0.5 mg Na_2SeO_3/L + 5 mg KIO_3./L	Phenolic compounds, phenylpropanoids, flavonols, and anthocyanins	[11]

Species	Plant part	Biofortification method	Bioactive compounds	Reference
Ocimum basilicum L.	Leaves	Foliar application: 0, 50, and 100 mg SeNPs/L	Monoterpenes, carotenoids, and phenolic acids	[37]
Capsicum annuum L.	Fruit	Foliar application every 15 d (6 times): 0, 1.5, 3.0, 4.5, and 6 mg Na_2SeO_3/L	Capsaicin, phenolic compounds, and total flavonoids	[39]
Brassica oleracea L.	Florets	Foliar application every 15 d: 0.1, 0.2, 0.4, 0.8, and 1.6 mg/L, Na_2SeO_3 and organic Se.	Glucosinolates, phenols, and total flavonoids	[42]
Brassica juncea L.	Leaves	Amended soil: 0, 2, 4, and 6 μM Na_2SeO_4/kg	Total phenols, flavonoids, and anthocyanins	[40]
Camellia sinensis L. Kuntze	Leaves	Daily foliar application for 5 d: 0 y 2 mg Na_2SeO_3/L d + low T (4 °C).	Caffeine, gallic acid, and flavonoids	[43]
Coriandrum sativum L., *Ocimum basilicum* L., *Brassica rapa* L. subsp. *narinosa*	Microgreens	Daily nutrient solution: 0, 1.5, and 3.0 mg Na_2SeO_4/L, 12 d coriander and 19 d basil.	Carotenoids and phenolic compounds	[41]
Brassica oleracea L.	Florets	Root irrigation every 15 d: 0, 17.3, 34.6, 69.2, 138.3, and 276.6 mg Na_2SeO_4/L	Glucosinolates and flavonoids	[44]
Brassica oleracea L.	Florets	Foliar application: 1, 5, 10, 15, and 20 mg Na_2SeO_4/L, three times (in weeks 2, 5, and 8).	Glucosinolates and phenolic compounds	[45]
Solanum lycopersicum L.	Fruit	Foliar application at the onset of flowering: 1 mg Na_2SeO_4/L	Lycopene and total flavonoids	[46]

Table 2.
Bioactive compounds induced in edible parts with selenium biofortification.

to prevent oxidative stress generated by low-temperature stress [43]. In contrast, Se + I and Se + I + AS (0.1, 1.0, and 10.0 mg/L) combinations did not induce changes in the total contents of phenolic compounds, flavonols, phenylpropanoids, and anthocyanins in lettuce leaves [11].

These results place biofortification with Se as a promising agronomic strategy for obtaining functional foods.

5. Health benefits from the intake of biofortified crops with Se

One of the most recognized biological activities of Se is its contribution to antioxidant processes, as well as its role as a chemopreventive agent since an adequate intake of Se can reduce the risk of cancer. In addition, many plant foods contain compounds

Food plant	Effect on human health	Suggested mechanism	Reference
Rice	Antioxidant	Increased glutathione (GSH), ascorbic acid (AsA), and glutathione peroxidase (GSH-Px) activity	[29]
Soybean sprouts	Antioxidant	Increased vitamin C and GSH content. Increases in peroxidase (POD) and ascorbate peroxidase (APX) activity	[34]
Chickpea sprouts	Antioxidant	Entry into cells (Caco-2) to combat oxidative stress	[47]
Green tea	Antioxidant	Increased superoxide dismutase (SOD) activity and epigallocatechin gallate content	[43]
Tomato	Antioxidant	Increased vitamin C, E, and GSH content	[46]
Lettuce	Anti-inflammatory	Inhibition of inducible nitric oxide synthase (iNOS) activity. Increased quercetin content	[6]
Coriander and tatsoi microgreens	Anti-inflammatory	Increased rutin and kaemferol-3-O-(feruloyl) sophoroside-7-O-glucoside content	[41]
Peanut	Anticarcinogenic	Inhibition in the proliferation of Caco-2 and HepG2 cell lines.	[32]
Soybean sprouts	Anticarcinogenic	Increased isoflavones content	[34]
Broccoli sprouts	Anticarcinogenic	Increased glucorapinin and glucoerucin content, precursors of anticancer compounds	[48]
Chickpea sprouts	Anticarcinogenic	Increased GSH-Px and thioredoxin reductase (TrxR) activities. Overexpression of Fas protein	[49]

Table 3.
Effect of Se-enriched plant-source foods on human health.

with important biological activities for disease control. Thus, research has shown that the biofortification of crops improves the antioxidant, anti-inflammatory, and anticarcinogenic properties of edible parts of plants (**Table 3**).

5.1 Antioxidant activity

In addition to providing organic Se species and bioactive compounds, biofortification with Se provides plant foods that benefit health through ingestion of edible parts with antioxidant capacity. Root application of selenite increased 85.9% glutathione (GSH) content, 39.2% ascorbic acid (AsA), and 186.0% glutathione peroxidase (GSH-Px) enzyme activity, indicating that Se biofortification increases the antioxidant capacity of rice grains [29]. Similarly, in soybean sprouts enriched with Se (selenite and SeNPs), an average 3-fold increase in vitamin C and 38% increase in GSH content were reported, as well as an increase in the activity of the antioxidant enzymes catalase (CAT), peroxidase (POD), superoxide dismutase (SOD), and ascorbate peroxidase (APX). Higher activity of POD (72–176% higher activity) and APX (2.5 times higher than the control) enzymes was highlighted in soybean sprouts enriched with 100 μM selenite and SeNPs [34]. Selenium treatments improved the antioxidant properties of soybean sprouts, so their consumption could improve human health. It is important to highlight the benefits of vitamin C, which is recognized as an excellent antioxidant that protects plants from ROS and plays a vital role in the human body. In another study, it was found that the protein fraction from chickpea sprouts, enriched with Se (2 mg selenite/100 g of seeds), had a significant increase in cellular antioxidant activity (CAA). The highest percentage of CAA was detected in peptides <10 kDa with Se supplementation (59.11 ± 2.06%), a CAA value equivalent to that of SeMet, SeCys, or selenite. The antioxidant activity assay indicated that Se species entered cells (Caco-2) at supranutritional doses, exerting different mechanisms to combat oxidative stress, those mainly related to redox cycles such as cell signaling, DNA stability, cell cycle genes and proliferation, reduction of the inflammatory response, caspases-mediated apoptosis, angiogenesis, and osteoclast inactivation [47]. The antioxidant activity of selenoproteins was a function of their Se content.

In green tea plants, induction of SOD activity and an increase in the content of epigallocatechin gallate (EGCG, 15.1%) and other catechins in response to Se application resulted in a reduction in the content of hydrogen peroxide (H_2O_2, 31.6%) and malondialdehyde (23.9%) [43]. The latter is a good indicator of lipid peroxidation. In addition to its antioxidant and chelating properties, EGCG has shown therapeutic potential as an anti-inflammatory, antibacterial, and antiviral, as well as for cancer prevention [50], which associates with numerous health benefits. In broccoli, the antioxidant capacity induced with Se biofortification depends on the cultivar, highlighting the 40% increase in Graffiti, while in the cultivar Clapton, only a 29% increase was recorded at the same concentration (5 mg Na_2SeO_4/L) [45]. Foliar application of Se also has an effect on antioxidant properties in tomato fruits, inducing vitamin C (1.3-fold higher than the control) and vitamin E (1.4-fold) production, as well as a 2-fold increase in reduced glutathione levels [46].

5.2 Anti-inflammatory activity

Se has been shown to have beneficial effects in the treatment of inflammatory diseases. Inflammation is characterized by the presence of pain, redness, swelling, and impaired function [6]. There are different markers that mediate immune cell recruitment and response to infection or injury. Among these, the enzyme inducible nitric oxide synthase (iNOS), responsible for the formation of nitric oxide (NO), plays an important role during the inflammation process. In this regard, Se was reported to modify the anti-inflammatory properties of lettuce plants that were grown under Se application, determined by inhibition of iNOS activity [6]. In addition, an increase in quercetin 3-O-(600-acetyl-glucoside) content was found. Quercetin and kaempferol are among the most common metabolites found in vegetables and fruits, which are considered to have high anti-inflammatory and antioxidant activity in in vitro studies [6]. It is important to note that biofortification with Se favors the synthesis of these compounds in different species, since in coriander and tatsoi microgreens, increases of 33 and 157% in rutin and kaemferol-3-O-(feruloyl) sophoroside-7-O-glucoside content are achieved at a concentration of 1.5 mg/L [41]. This induction of Se is carried out at the transcriptional level in broccoli, favoring the expression of genes of the phenylpropanoid pathway [44]. Caffeic acid is another secondary metabolite that is increased in lettuce plants biofortified with Se nanomaterials and is considered one of the bioactive compounds of propolis with antitumor and anti-inflammatory effects [38].

5.3 Anticarcinogenic activity

It has already been mentioned that one way to include Se in the human diet is through the consumption of cultivated plants enriched with Se. Therefore, Se intake is of vital importance both to cover nutritional demand and for the prevention of health problems such as cancer. A study evaluated the anticancer activity of Se-enriched peanut protein and found an inhibitory effect on Caco-2 and HepG2 cell lines. Furthermore, it was reported that peanut protein, obtained from Se biofortification, significantly inhibited cell proliferation in a dose-dependent manner, with a dose range of 15.6 to 250 μg/mL, with the 250 μg/mL dose being more effective [32]. These studies provide solid information on the anticancer effect of Se-enriched peanut protein.

Secondary metabolites such as isoflavones have also been reported to have bioactivity for cancer prevention and treatment. In this regard, Rao et al. [34] found that selenite and SeNPs promoted the accumulation of total isoflavones in soybean sprouts. In addition to tasting good for direct and fresh consumption, soybean sprouts contain health-promoting substances such as vitamin C and isoflavones.

Another group of phytochemical compounds with important anticarcinogenic activity is glucosinolates. Brassicas are crops recognized as chemopreventive foods. Therefore, biofortification with Se during Brassica cultivation would be expected to increase the chemopreventive activity of the sprouts. In sprouts of three broccoli cultivars, enriched with selenate, glucoraphanin was found to be the dominant glucosinolate, accounting for 70% of the total glucosinolate content. Glucoraphanin is an aliphatic glucosinolate and is a direct precursor of sulforaphane isothiocyanate, which acts as a potent monoinducer of phase II-related enzymes during the inactivation of carcinogenic metabolites. Another aliphatic glucosinolate present in broccoli sprouts is glucoerucin, accounting for 14% of the total glucosinolate content, which is metabolized to the isothiocyanate erucin, considered an anticarcinogenic agent [48]. Therefore, broccoli sprouts could be considered an excellent source for the intake of isothiocyanate compounds for cancer prevention.

Se-enriched chickpea sprouts were found to be an important source of dietary Se and isoflavonoids with chemopreventive potential for the treatment of colorectal cancer. A diet enriched with a supranutritional dose of Se (2.29 μg/g diet) in combination with isoflavonoids (2.34 mg/g) was tested on tumor growth of xenoplastic human colorectal adenocarcinoma cells in immunosuppressed mice [49]. The diet promoted cell apoptosis through overexpression of cell surface death receptor (Fas). In addition, an increase in GSH-Px and thioredoxin reductase (TrxR) enzyme activity was observed; as well as an increase in cholesterol, triglycerides, and low-density lipoprotein cholesterol, resulting in a significant decrease in tumor cell growth [49]. These types of studies indicate that ingestion of chickpea sprouts enriched with Se can contribute to reducing cancer cell proliferation.

6. Conclusions

Agronomic biofortification is becoming the most widely used strategy for Se supplementation of plant foods because it is a relatively simple agronomic practice to operate and because of its high availability. The distribution of organic and inorganic Se species is a key factor to consider in the biofortification process. There are a large number of cultivated plants that have the ability to convert inorganic Se (mainly selenate or selenite) into organic Se (SeCys, SeMet, or MetSeCyt), representing an excellent metabolic mechanism for obtaining Se-rich foods.

Selenium applied in different forms enhances the accumulation of phytochemicals with antioxidant, anti-inflammatory, antimicrobial, and antitumor properties in different edible plant species, highlighting the advantages of incorporating biofortification with Se in the production chain of foods rich in bioactive compounds, which is a desirable feature in the food industry due to the positive impact on human health. Therefore, it is imperative to elucidate the mechanisms by which this trace element induces the production of these biocompounds in plants in order to optimize this strategy.

Acknowledgements

The authors would like to thank the National Council of Science and Technology (CONACYT, Mexico) for the M. Sc. scholarship of Luis Guillermo Terreros-Rosales (1117428). SG-M is a PLANTECC National Laboratory member.

Conflict of interest

The authors declare no conflict of interest.

Author details

Soledad García-Morales[1*], Janet María León-Morales[2], Víctor García-Gaytán[3] and Luis Guillermo Terreros-Rosales[4]

1 CONACYT-Center for Research and Assistance in Technology and Design of the State of Jalisco, Zapopan, Mexico

2 Academic Coordination West Altiplano Campus-Autonomous University of San Luis Potosi, Salinas de Hidalgo, Mexico

3 Plant Nutrition and Horticulture, Mexico

4 Center for Research and Assistance in Technology and Design of the State of Jalisco, Zapopan, Mexico

*Address all correspondence to: smorales@ciatej.mx

References

[1] Guan BZ, Yan RL, Li RM, Zhang X. Selenium as a pleiotropic agent for medical discovery and drug delivery. International Journal of Nanomedicine. 2018;**13**:7473-7490. DOI:10.2147/IJN.S181343

[2] Leonardi A, Evke S, Lee M, Melendez JA, Begley TJ. Epitranscriptomic systems regulate the translation of reactive oxygen species detoxifying and disease linked selenoproteins. Free Radical Biology and Medicine. 2019;**143**:573-593. DOI:10.1016/j.freeradbiomed.2019.08.030

[3] Constantinescu-Aruxandei D, Frîncu RM, Capră L, Oancea F. Selenium analysis and speciation in dietary supplements based on next-generation selenium ingredients. Nutrients. 2018;**10**:1466. DOI:10.3390/nu10101466

[4] Garza-García JJO, Hernández-Díaz JA, Zamudio-Ojeda A, León-Morales JM, Guerrero-Guzmán A, Sánchez-Chiprés DR, et al. The role of selenium nanoparticles in agriculture and food technology. Biological Trace Element Research. 2022;**200**:2528-2548. DOI:10.1007/s12011-021-02847-3

[5] Niedzielski P, Rudnicka M, Wachelka M, Kozak L, Rzany M, Wozniak M, et al. Selenium species in selenium fortified dietary supplements. Food Chemistry. 2016;**190**:454-459. DOI:10.1016/j.foodchem.2015.05.125

[6] Abdalla MA, Famuyide I, Wooding M, McGaw LJ, Mühling KH. Secondary metabolite profile and pharmacological opportunities of lettuce plants following selenium and Sulfur enhancement. Pharmaceutics. 2022;**14**:2267. DOI:10.3390/pharmaceutics14112267

[7] Szerement J, Szatanik-Kloc A, Mokrzycki J, Mierzwa-Hersztek M. Agronomic biofortification with Se, Zn, and Fe: An effective strategy to enhance crop nutritional quality and stress defense—A review. Journal of Soil Science and Plant Nutrition. 2022;**22**:1129-1159. DOI:10.1007/s42729-021-00719-2

[8] Li Y, Xiao Y, Hao J, Fan S, Dong R, Zeng H, et al. Effects of selenate and selenite on selenium accumulation and speciation in lettuce. Plant Physiology and Biochemistry. 2022;**192**:162-171. DOI:10.1016/j.plaphy.2022.10.007

[9] Zhang H, Zhao Z, Zhang X, Zhang W, Huang L, Zhang Z. Effects of foliar application of selenate and selenite at different growth stages on selenium accumulation and speciation in potato (*Solanum tuberosum* L). Food Chemistry. 2019;**286**:550-556. DOI:10.1016/j.foodchem.2019.01.185

[10] Wang M, Zhou F, Cheng N, Chen P, Ma Y, Zhai H, et al. Soil and foliar selenium application: Impact on accumulation, speciation, and bioaccessibility of selenium in wheat (*Triticum aestivum* L.). Frontiers in Plant Science. 2022;**13**:988627. DOI:10.3389/fpls.2022.988627

[11] Smolen S, Kowalska I, Czernicka M, Halka M, Kęska K, Sady W. Iodine and selenium biofortification with additional application of salicylic acid affects yield, selected molecular parameters and chemical composition of lettuce plants (*Lactuca sativa* L. var. capitata). Frontiers. Plant Science. 2016;**7**:1553. DOI:10.3389/fpls.2016.01553

[12] Smolen S, Kowalska I, Skoczylas L, Liszka-Skoczylas M, Grzanka M, Halka M,

et al. The effect of salicylic acid on biofortificaton with iodine and selenium and the quality of potato cultivates in the NFT system. Scientia Horticulturae. 2018;**240**:530-543. DOI:10.1016/j.scienta.2018.06.060

[13] Skrypnik L, Styran T, Savina T, Golubkina N. Effect of selenium application and growth stage at harvest on hydrophilic and lipophilic antioxidants in lamb's lettuce (*Valerianella locusta* L. Laterr). Plants. 2021;**10**(12):2733. DOI:10.3390/plants10122733

[14] Perez-Millan R, Alfosea-Simon M, Zavala-Gonzalez EA, Camara-Zapata JM, Martinez-Nicolas JJ, Lidon V, et al. The addition of selenium to the nutrient solution decreases cadmium toxicity in pepper plants grown under hydroponic conditions. Agronomy. 2021;**11**(10):1905. DOI:10.3390/agronomy11101905

[15] Pérez-Millán R, Cámara-Zapata JM, Fernández-Zapata JC, Simón-Grao S, Alfosea-Simón M, Zavala-González EA, et al. Application of biocat G, selenium, and chitosan to counteract the negative effects of Cd in broccoli plants grown in soilless culture. Agronomy. 2022;**12**(6):1327. DOI:10.3390/agronomy12061327

[16] Lanza MGDB, Dos Reis AR. Roles of selenium in mineral plant nutrition: ROS scavenging responses against abiotic stresses. Plant Physiology and Biochemistry. 2021;**164**:27-43. DOI:10.1016/j.plaphy.2021.04.026

[17] Golubkina N, Moldovan A, Kekina H, Kharchenko V, Sekara A, Vasileva V, et al. Joint biofortification of plants with selenium and iodine: New field of discoveries. Plants. 2021;**10**(7):1352. DOI:10.3390/plants10071352

[18] Zou C, Du Y, Rashid A, Ram H, Savasli E, Pieterse PJ, et al. Simultaneous biofortification of wheat with zinc, iodine, selenium, and iron through foliar treatment of a micronutrient cocktail in six countries. Journal of Agricultural and Food Chemistry. 2019;**29**:8096-8106. DOI:10.1021/acs.jafc.9b01829

[19] Prom-U-Thai C, Rashid A, Ram H, Zou C, Guilherme LRG, Corguinha APB, et al. Simultaneous biofortification of rice with zinc, iodine, iron and selenium through foliar treatment of a micronutrient cocktail in five countries. Frontiers. Plant Science. 2020;**11**:589835. DOI:10.3389/fpls.2020.589835

[20] Wang M, Ali F, Qi M, Peng Q, Wang M, Bañuelos GS, et al. Insights into uptake, accumulation, and subcellular distribution of selenium among eight wheat (*Triticum aestivum* L.) cultivars supplied with selenite and selenate. Ecotoxicology and Environmental Safety. 2021;**207**:111544. DOI:10.1016/j.ecoenv.2020.111544

[21] Gupta M, Gupta S. An overview of selenium uptake, metabolism, and toxicity in plants. Frontiers in Plant Science. 2017;7:2074. DOI:10.3389/fpls.2016.02074

[22] Zhou X, Yang J, Kronzucker HJ, Shi W. Selenium biofortification and interaction with other elements in plants: A review. Frontiers in Plant Science. 2020;**11**:586421. DOI:10.3389/fpls.2020.586421

[23] García-Gaytán V, García-Morales S, Coria-Téllez AV, Hernández-Mendoza F, León-Morales JM. *Solanum lycopersicum*: Novelties for climate factors and its agronomic management for high yields. World Journal of Engineering and Technology. 2022;**3**:605-612. DOI:10.4236/wjet.2022.103039

[24] Siddiqui SA, Blinov AV, Serov AV, Gvozdenko AA, Kravtsov AA,

Nagdalian AA, et al. Effect of selenium nanoparticles on germination of *Hordéum Vulgáre* barley seeds. Coatings. 2021;**11**:862. DOI:10.3390/coatings11070862

[25] Amin MA, Ismail MA, Badawy AA, Awad MA, Hamza MF, Awad MF, et al. The potency of fungal-fabricated selenium nanoparticles to improve the growth performance of *Helianthus annuus* L. and control of cutworm Agrotis ipsilon. Catalysts. 2021;**11**:1551. DOI:10.3390/catal11121551

[26] Hashem AH, Abdelaziz AM, Attia MS, Salem SS. Selenium and Nano-Selenium-Mediated Biotic Stress Tolerance in Plants. In: Hossain MA, Ahammed GJ, Kolbert Z, El-Ramady H, Islam T, Schiavon M, editors. Selenium and Nano-Selenium in Environmental Stress Management and Crop Quality Improvement. Cham: Springer; 2022. pp. 209-226. DOI:10.1007/978-3-031-07063-1_11

[27] Xu X, Wang J, Wu H, Yuan Q, Wang J, Cui J, et al. Effects of selenium fertilizer application and tomato varieties on tomato fruit quality: A meta-analysis. Scientia Horticulturae. 2022;**304**:111242. DOI:10.1016/j.scienta.2022.111242

[28] Naz M, Shabbir R, Verma KK, Rastogi A, Rajput VD, Javed T, et al. Recent developments to mitigate selenium deficiency in agricultural eco-systems. Phyton. 2022;**915**:916-927. DOI:10.32604/phyton.2022.018688

[29] Yin H, Qi Z, Li M, Ahammed GJ, Chu X, Zhou J. Selenium forms and methods of application differentially modulate plant growth, photosynthesis, stress tolerance, selenium content and speciation in *Oryza sativa* L. Ecotoxicology and Environmental Safety. 2019;**169**:911-917. DOI:10.1016/j.ecoenv.2018.11.080

[30] Hu Z, Cheng Y, Suzuki N, Guo X, Xiong H, Ogra Y. Speciation of selenium in Brown Rice fertilized with selenite and effects of selenium fertilization on Rice proteins. International Journal of Molecular Sciences. 2018;**19**:3494. DOI:10.3390/ijms19113494

[31] Muleya M, Young SD, Vazquez RS, Ligowe IS, Broadley MR, Joy EJM, et al. Selenium speciation and bioaccessibility in Se-fertilised crops of dietary importance in Malawi. Journal of Food Composition and Analysis. 2021;**98**:103841. DOI:10.1016/j.jfca.2021.103841

[32] Luo L, Zhang J, Zhang K, Wen Q, Ming K, Xiong H, et al. Peanut selenium distribution, concentration, speciation, and effects on proteins after exogenous selenium biofortification. Food Chemistry. 2021;**354**:129515. DOI:10.1016/j.foodchem.2021.129515

[33] Sánchez-Rodas D, Mellano F, Martínez F, Palencia P, Giráldez I, Morales E. Speciation analysis of Se-enriched strawberries (*Fragaria ananassa Duch*) cultivated on hydroponics by HPLC-TR-HG-AFS. Microchemical Journal. 2016;**127**:120-124. DOI:10.1016/j.microc.2016.02.017

[34] Rao S, Xiao X, Wang Y, Xiong Y, Cheng H, Li L, et al. Comparative study of the effects of selenium nanoparticles and selenite on selenium content and nutrient quality in soybean sprouts. Folia Horticulturae. 2022;**34**:223-234. DOI:10.2478/fhort-2022-0017

[35] Zhang Y-J, Gan R-Y, Li S, Zhou Y, Li A-N, Xu D-P, et al. Antioxidant phytochemicals for the prevention and treatment of chronic diseases. Molecules. 2015;**20**:21138-21156. DOI:10.3390/molecules201219753

[36] García-Morales S, Macías-García MJ, Lugo-Cervantes E, Alcázar-Valle EM.

Effects of selenium on yield, seed size, and phenolic compound content of common bean (*Phaseolus vulgaris* L.). In: Marroquín A, Castillo L, Soto S, Cruz L, coordinators. Handbook T-XIX, CIERMMI Women in Science, Biological Sciences. Queretaro: ©ECORFAN-Mexico; 2022. p. 82-95. DOI:10.35429/H.2022.6.1.82.95

[37] Asghari J, Mahdavikia H, Rezaei-Chiyaneh E, Banaei-Asl F, Machiani MA, Harrison MT. Selenium nanoparticles improve physiological and phytochemical properties of basil (*Ocimum basilicum* L.) under drought stress conditions. Land. 2023;**12**:164. DOI:10.3390/land12010164

[38] Cheng B, Wang C, Yue L, Chen F, Cao X, Lan Q, et al. Selenium nanomaterials improve the quality of lettuce (*Lactuca sativa* L.) by modulating root growth, nutrient availability, and photosynthesis. NanoImpact. 2023;**29**:100449. DOI:10.1016/j.impact.2022.100449

[39] Gaucin-Delgado JM, Preciado-Rangel P, González-Salas U, Sifuentes-Ibarra E, Núñez-Ramírez F, Orozco Vidal JA. La biofortificación con selenio mejora los compuestos bioactivos y la actividad antioxidante en chile jalapeño. Revista Mexicana de Ciencias Agrícolas. 2021;**12**:1339-1349. DOI:10.29312/remexca.v12i8.3066

[40] Handa N, Kohli SK, Sharma A, Thukral AK, Bhardwaj R, Abdllah EF, et al. Selenium modulates dynamics of antioxidative defence expression, photosynthetic attributes and secondary metabolites to mitigate chromium toxicity in *Brassica juncea* L. plants. Environmental and Experimental Botany. 2019;**161**:180-192. DOI:10.1016/j.envexpbot.2018.11.009

[41] Pannico A, El-Nakhel C, Graziani G, Kyriacou MC, Giordano M, Soteriou GA, et al. Selenium biofortification impacts the nutritive value, polyphenolic content, and bioactive constitution of variable microgreens genotypes. Antioxidants. 2020;**9**:272. DOI:10.3390/antiox9040272

[42] Gui J-Y, Rao S, Gou Y, Xua F, Cheng S. Comparative study of the effects of selenium yeast and sodium selenite on selenium content and nutrient quality in broccoli florets (*Brassica oleracea* L. var. italica). Journal of the Science of Food and Agriculture. 2022;**102**:1707-1718. DOI:10.1002/jsfa.11511

[43] Liu K, Li S, Han J, Zeng X, Ling M, Mao J, et al. Effect of selenium on tea (*Camellia sinensis*) under low temperature: Changes in physiological and biochemical responses and quality. Environmental and Experimental Botany. 2021;**188**:104475. DOI:10.1016/j.envexpbot.2021.104475

[44] Rao S, Gou Y, Yu T, Cong X, Gui J, Zhu Z, et al. Effects of selenate on Se, flavonoid, and glucosinolate in broccoli florets by combined transcriptome and metabolome analyses. Food Research International. 2021;**146**:110463. DOI:10.1016/j.foodres.2021.110463

[45] Saeedi M, Soltani F, Babalar M, Izadpanah F, Wiesner-Reinhold M, Baldermann S. Selenium fortification alters the growth, antioxidant characteristics and secondary metabolite profiles of cauliflower (*Brassica oleracea* var. botrytis) cultivars in hydroponic culture. Plants. 2021;**10**:1537. DOI:10.3390/plants10081537

[46] Zhu Z, Zhang Y, Liu J, Chen Y, Zhang X. Exploring the effects of selenium treatment on the nutritional quality of tomato fruit. Food Chemistry. 2018;**252**:9-15. DOI:10.1016/j.foodchem.2018.01.064

[47] Serrano-Sandoval SN, Guardado-Félix D, Gutiérrez-Uribe JA. Changes

in digestibility of proteins from chickpeas (Cicer arietinum L.) germinated in presence of selenium and antioxidant capacity of hydrolysates. Food Chemistry. 2019;**285**:290-295. DOI:10.1016/j.foodchem.2019.01.137

[48] Ávila FW, Yang Y, Faquin V, Ramos SJ, Guilherme LRG, Thannhauser TW, et al. Impact of selenium supply on Se-methylselenocysteine and glucosinolate accumulation in selenium-biofortified Brassica sprouts. Food Chemistry. 2014;**165**:578-586. DOI:10.1016/j.foodchem.2014.05.134

[49] Guardado-Félix D, Antunes-Ricardo M, Rocha-Pizaña MR, Martínez-Torres AC, Gutiérrez-Uribe JA, Serna Saldivar SO. Chickpea (*Cicer arietinum* L.) sprouts containing supranutritional levels of selenium decrease tumor growth of colon cancer cells xenografted in immune-suppressed mice. Journal of Functional Foods. 2019;**53**:76-84

[50] Granja A, Frias I, Neves AR, Pinheiro M, Reis S. Therapeutic potential of epigallocatechin gallate nanodelivery systems. Biomed Research International. 2017;**2017**:5813793. DOI:10.1155/2017/5813793

Chapter 9

Distribution of Selenium in Soils and Human Health

Muhammad Imran, Zhikun Chen, Ayaz Mehmood, Shah Rukh, Wang Weixie, Waleed Asghar and Farhan Iftikhar

Abstract

Selenium (Se) is essential as well as a toxic element for humans and animals if it exceeds a certain limit. Soil selenium plays an important role through the food chain. Total selenium in world soils ranges from 0.125 to 0.3 mg kg^{-1} and varies with the soils' nature. High and low selenium in soils also poses serious environmental and health risks. However, in addition to selenium's overall quantity in soil, selenium reactivity, and bioavailability also depend on its chemical structure. The amount of available selenium in the soil varies depending on its oxidation state since selenium species include selenide (Se^{2-}), elemental selenium (Se^0), selenite (Seo_3^{2-}), selenate (Seo_4^{2-}). The pH, soil texture, amount of organic matter, and the presence of competing ions are the four most significant soil characteristics that affect Se availability. Similarly, selenium uptake and accumulation are influenced by the crop type whether it is an accumulator or not. The selenium environmental and health risk assessment is necessary to evaluate in soils with high selenium contents and crops with higher selenium uptake. Whereas in areas where selenium deficiency is observed or vulnerable to selenium, deficiency needs to be supplemented through Se inputs. The selenium deficiency and toxicity areas should be monitored carefully from a health perspective.

Keywords: total soil selenium, selenium species, selenium transformation, selenium bioavailability, selenium risk assessment

1. Introduction

A vital element for both humans and animals, selenium (Se) is a metalloid that lies in the middle of the metal and non-metal. Selenium plays a crucial role in the biological processes of human and animals body. Its high concentrations make it poisonous, and a lack of it can have catastrophic consequences on human and animal health [1]. Despite the fact that selenium has a wide range of important advantages, selenium insufficiency is becoming a widespread issue around the world. A health danger exists when selenium intake is excessive. Moreover, type II diabetes risk may be increased by a diet high in selenium [2]. A high selenium intake may enhance the expression of the transcription coactivator peroxisome proliferator-activated receptor-coactivator

(PGC-1), which is important in cellular energy metabolism and may result in hyperglycemia [3]. Excessive selenium consumption results in loss of hair and nails, damage to the neurological system, paralysis, and even death [4]. The daily selenium consumption dosage so has significance. Selenium 40 μg d^{-1} the recommended daily allowance (RDA) suggested by the WHO [5]. An overdose occurs when the consumption for men is greater than 60 μg d^{-1} and for women is greater than 53 μg d^{-1} [6]. Responses varied when referring to various forms of selenium, and this is for the total amount of selenium.

Selenium availability from soil affects the food chain selenium level. Three major selenium mineral i.e. tiemannite (HgSe), clausthalite (PbSe), and naumannite ((Ag, Pb) Se) contains selenium and is present in soils [7]. The soil's total selenium depends on the type of parent materials and the soil-forming processes which redistribute selenium [8]. Overall, total selenium in world soils ranges from 0.125 to 0.3 mg kg^{-1} and varies with the soils' nature [9]. The essential level of selenium for animals ranges from 0.04 to 0.1 mg kg^{-1}, while a concentration exceeding 3.5 to 5 mg kg^{-1} in their food may cause harmful impacts [4, 10]. Human activities including fossil fuel and coal burning, metal smelting, inorganic, and organic fertilizer application, lime, manure, and solid sewage waste disposal cause Se accumulation in soils [11]. The selenium accumulation in soils poses serious threats to the agroecosystem via bioaccumulation [11, 12]. Selenium toxicity in soil and food chains depends on its forms and distribution rather than its total contents [13]. Total selenium concentration in soils derived from various sources ranges from 0.27 to 7.05 mg kg^{-1} [8].

Yet in addition to the amount of selenium in the soil as a whole, selenium reactivity and bioavailability also depend on the chemical form of the element. Several forms of selenium, including selenide, elemental selenium, selenite, selenate, and organic selenium, are found in soil, depending on its oxidation state [14]. The replenishment of selenium in soil solution is also aided by selenium that is contained in or bonded to various fractions in soils. Typically, there are five different selenium fractions: ion-exchangeable or calcium-bound selenium, oxides-bound selenium (iron and aluminum oxides), organic and humic-bound selenium, sulfide-bound selenium, and residual selenium. Thus, it's critical to keep an eye on the type and amount of selenium exposure through different foods grown under different soils.

2. Selenium species in soils

There are a variety of selenium species that can be found in soil solution.

$$\text{Selenate (Se VI)} = \left(\text{Seo}_4^{2-}\right), \left(\text{HSeo}_4^{-}\right), \left(\text{H}_2\text{Seo}_4^{-}\right)$$

$$\text{Selenite (SeIV)} = \left(\text{Seo}_3^{2-}\right), \left(\text{HSeo}_3^{-}\right), \left(\text{H}_2\text{Seo}_3^{0}\right)$$

$$\text{Selenate(II)} = \left(\text{Se}^{2-}\right), \left(\text{HSe}^{-}\right), \left(\text{H}_2\text{Se}^{0}\right)$$

Depending on the characteristics of the environment or the soil, several species of selenium can be found in the form of selenide (Se^{2-}), elemental selenium (Se^0), selenite (**Seo_3^{2-}**), selenate (**Seo_4^{2-}**). Selenate (**Seo_4^{2-}**): Under conditions in which it is thoroughly oxidized, selenate maintains its stability. Selenate is not absorbed by soil elements with the same level of strength as selenite [15, 16], and the transformation of selenate into less mobile forms of selenite or elemental Se) is a long process [17].

Selenate is the form of selenium that may be taken up by plants in the greatest quantity [18, 19]. Selenite (**Seo_3^{2-}**): Selenite is a can be found in settings that are only slightly oxidized. Selenous acid is a weak acid that can only be protonated in conditions where the pH values range from acidic to neutral. Microorganisms in acidic settings [20] or moderately reducing agents in neutral or alkaline environments [21] can convert selenite to elemental selenium. Selenite possesses a significant propensity for sorption, in particular by oxides of iron and aluminum [17, 18]. Whereas the adsorption of selenite (**Seo_3^{2-}**) depends on pH, and the concentration of competing anions such as phosphate (**PO_4^{3-}**) [22]. Selenide: Selenide (Se^{2-}) typically exists in reducing environments as metal selenides and hydrogen selenide (H_2Se) a poisonous gas with a bad smell. In water, it readily oxidizes to elemental Se [23]. Se-sulfides and metal selenides often have very low solubility [24]. Besides that, microbial activities also result in the production of dissolved organic selenide molecules or volatile methylated derivatives of selenium such as dimethyl diselenide [25, 26]. Elemental Selenium (Se^0): Elemental selenium (Se^0) exists in reduced conditions in the form of crystalline or amorphous. Red crystalline Se is alpha- and beta-monoclinic Se. Whereas the amorphous form is Red and glassy or black [27]. Elemental Se oxidize or reduce slowly and extremely insoluble in water. Specific microorganisms can oxidize elemental Se to selenite (**Seo_3^{2-}**), and selenate (**Seo_4^{2-}**) [23].

3. Selenium solubility and transformation

Selenate and selenite are the major forms of Se in cultivated soils. The mole fractions of Se species were used to calculate the total soluble Se supported by eight selenate and selenite minerals, which might be present in soils. The effect of redox on total soluble Se at which these minerals can form in neutral soils. None of these minerals are expected to form in normally cultivated soils. Only manganese selenite ($MnSeO_3$) is sufficiently stable that it might precipitate in strongly acidic environments. Ferric selenite was included because several investigators reported that it might be formed in acid soils. Decreasing pH has a negative effect on the solubility of both minerals which also suggests that $Fe_2(SeO_3)_3$ is unstable with respect to $MnSeO_3$. At pH 4, $Fe_2(SeO_3)_3$ and $MnSeo_3$ can maintain $10^{-1.5}$ and $10^{-6.7}$ M of Se in the solution. The previous studies' reported for soluble Se in acid soils appeared to be close to the solubility of $MnSeO_3$. Drastic changes in pH have strong effects on precipitation/dissolution and adsorption/desorption processes in soils, and disturbed soil systems may need much longer time than pure systems to re-attain equilibrium. The concept that Se in soils is governed by an adsorption type of mechanism rather than by precipitation/dissolution reactions is accepted by most soil scientists. The sorption of Se in acid soils was related to sesquioxides.

The majority of the Se in agricultural soils exists as selenate or selenite. The sum of soluble Se can be calculated through the Se supplied through selenate and selenite minerals that could be calculated in soils by using the mole fractions of Se species. Changes in the amount of total soluble Se that are necessary for mineralization in neutral soils as a result of redox conditions. In typical agricultural soils, none of these minerals would be expected to occur. However, only manganese selenite ($MnSeO_3$) is stable enough to possibly precipitate in highly acidic conditions. Many researchers suggested that ferric selenite could be generated in acid soils, so it was included. Both minerals become less soluble as pH decreases, which is more evidence that $Fe_2(SeO_3)_3$

is more unstable than $MnSeO_3$ in acidic conditions. The compounds $Fe_2(SeO_3)_3$ and $MnSeO_3$ supply 101.5 and 106.7 M of Se, respectively, in soil solution at 4 pH. It appeared that the solubility of $MnSeO_3$ in acid soils was close to that reported in prior investigations. It may take significantly longer for disturbed soil systems to re-attain than pure systems, as large shifts in pH have profound effects on precipitation/dissolution and adsorption/desorption processes in soils. Most soil scientists agree that Se in soils is controlled by an adsorption mechanism rather than precipitation/dissolution events and sesquioxides are considered to play a significant role in the sorption of Se in acid soils.

Soil selenium can be found in a variety of oxidation states, including 2, 4, and 6. The chemical speciation and environmental stability of selenium compounds are largely controlled by redox potential and pH. The selenate species predominates throughout a wide pH range at high redox. At the middle of the redox scale, biselenite or selenite dominates depending on the pH. We anticipate the presence of elemental Se and selenide species only at low redox. The amount of Se in the liquid phase of acid soils may be regulated by adsorption and desorption processes. The chemical forms of an element in soil are regulated by the redox potential (Eh) and pH. It has been shown **Figure 1** [28] that when elemental Se is given to soils, some of it is rapidly oxidized to selenite, and that the rate of transfer from selenite to selenate and selenate to elemental Se is considerably slower. There was no correlation between soil pH and the rate of oxidation of elemental Se, but this oxidation rate did vary. In alkaline soils, selenite can be easily oxidized to selenate, while in acid soils, this process can be somewhat challenging [16]. Of all the Se oxides, selenium dioxide has the highest degree of

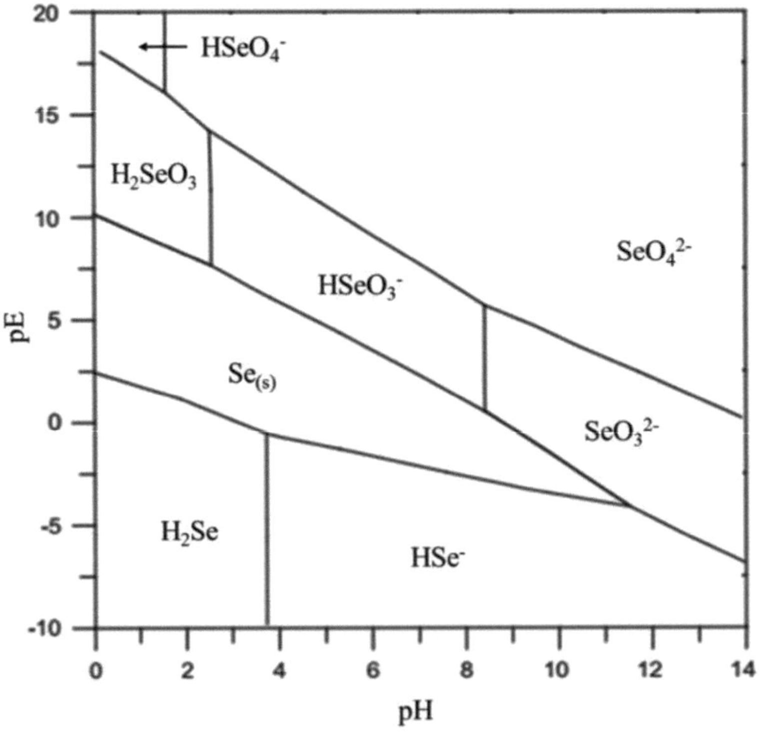

Figure 1.
Selenium forms at different pE-pH levels (adopted from Seby et al. 2001).

stability. Selenium dioxide is readily reduced to elemental Se by mild reducing agents [29]. Easily dissolving in water, selenium dioxide reacts with hydrogen peroxide to produce a weak acid called selenious acid. With the help of mild reducing agents like sulfur dioxide, selenite can be quickly converted to elemental Se in acidic circumstances [25]. In dry, alkaline conditions, elemental Se persists in the form of sandstones. it has been claimed that some soils contain elemental Se because they contain bacteria and fungi that can break down selenite and selenate into elemental Se. In addition, bacteria are capable of converting elemental Se into selenite or selenate under the right conditions.

4. Selenium bioavailability

The subject of selenium (Se) uptake by plants always remained an important aspect to study, because of its direct relation to human and animal health through the food chain. The high Se concentrations in food cause adverse health effects for humans [10]. While Se plays a crucial part in a variety of biological processes, which makes it a necessary trace element for both humans and other animals. Se insufficiency has been associated with multiple ailments, both in humans and in livestock. Due to the fact that Se is both an important nutrient and an element that can be poisonous, a substantial amount of study has been done to both enhance and decrease the amount of se that is present in plants.

The selenium uptake and accumulation by plants is a major concern because of its deficiency or toxicity through crops. To meet human and animal nutritional requirements it is very important to carefully consider the soil and crop factors to understand how uptake and accumulation in crop plants are influenced. In Se deficient regions how to increase Se concentrations in plant tissue and ultimately in the food chain. Whereas the Se-rich areas or seleniferous soils how can reduce the selenium uptake or accumulation in plants. Plant species also vary in Se accumulation. Unlike normal agricultural crops, which accumulate very little selenium, selenium accumulator plants can accumulate exceptionally high quantities of Se when cultivated in seleniferous soils [18, 23]. Yet the essential role of se for plants is not known. Plants uptake selenate (Se^{6+}) many folds greater than selenite (Se^{4+}). whereas the elemental Se (Se^{0}) is difficult or impossible for plants to obtain. Plants' ability to absorb selenium is also impacted by the chemical and physical properties of the soil, including pH level, soil texture, amount of organic matter, and the presence of ions like PO_4- and SO_4-. The concentration of selenium (Se) in soils and plants that are poor in Se can be increased by applying selenium (Se) to the soil, the seed, and the plant leaves.

4.1 Influencing factors on selenium bioavailability

The presence of competitive ions, electrical conductivity (Eh), pH, soil texture, and organic matter content are the five most critical soil variables that influence the availability of selenium. *The Eh and pH*: Both soil Eh and pH play a significant role in determining the chemical form that selenium takes up in soils. Selenate (Se^{6+}) is the predominant form of selenium (Se) found in well-aerated, alkaline soils. Selenite (Se^{4+}) is the predominant form of selenium in neutral and acid soils. Due to its adsorption by clays and iron oxides, selenite is slightly less readily available than other forms of selenium. The oxidation state of selenium is affected by the pH of the soil (**Figure 1**), but the ability of clays and ferric oxide to adsorb selenium is also impacted by this property [30]. Between pH 3 and 8, there was hardly any change in the amount of selenite (Se^{4+}) that was adsorbed by Fe_2O_3. They came to the conclusion that the pH, and not the layer silicate structure, was the factor that governed selenite (Se^{4+}) adsorption on clay minerals. The effects of pH on the effect on the sequestration of selenium by plants was also observed. The greatest quantity of selenite (Se^{4+}) is available to plants when those plants are cultivated on soils with a pH range of acidic to neutral. As the pH of the soil rises, hydroxyl ions take the place of selenite (Se^{4+}) on the adsorption sites. This causes selenite (Se^{4+}) to be released into the solution, which results in an increase in the availability of the element to plants [10]. *Soil texture*: Because selenite (Se^{4+}) is absorbed by clays, the proportion of clay in the soil has a significant bearing on how well plants are able to take it up. Hence, plants are able to absorb twice as much Se from sandy-textured soil. *Organic matter*: Selenium is released and fixed in part by organic materials. Organometallic complexes may offer significant Se-adsorbing sites, and organic matter fixes the selenium by removing it from the soil solution. Because organic matter in soil serves as a source of selenium, plants absorb more of it than they would in inorganic soils.

5. Accumulation and ecological risk assessment

Selenium accumulation was quantified by calculating the index of Se accumulation (Igeo). The geo-accumulation index was first purposed by Müller [31], to investigate heavy metals pollution compared with their background concentration in respective soils [32], it can be defined as follows:

$$Geo - accumulationIndex\ (I_{geo}) = log_2\left(\frac{C_{Soil}^{Se}}{k \times C_b^{Se}}\right) \quad (1)$$

In the above equation, C_{Soil}^{Se} denotes the selenium contents in soils and C_b^{Se} denotes the background concentration of selenium in respective soils. Whereas, in background concentration, the k is constant and its value is 1.5. The quantification of selenium contamination in soils was classified by geo-accumulation index criteria (**Table 1**).

The selenium pollution load index (PLI) was calculated as

Geo-accumulation index		Pollution load index		Ecological risk index	
Igeo	**Level**	**PLI**	**Level**	**ER**	**Level**
Igeo < 0	Uncontaminated	PLI ≤ 1	Low level of pollution	ER < 40	Low potential ecological risk
0 < Igeo ≤ 1	Uncontaminated to moderately contaminated	1 < PLI ≤ 2	Moderate level of pollution	40 ≤ ER < 80	Moderate potential ecological risk
1 < Igeo ≤ 2	Moderately contaminated	2 < PLI ≤ 5	High level of pollution	80 ≤ ER < 160	Considerable potential ecological risk
2 < Igeo ≤ 3	Moderately to heavily contaminated	PLI > 5	Extremely high level of pollution	160 ≤ ER < 320	High potential ecological risk

Geo-accumulation index		Pollution load index		Ecological risk index	
Igeo	**Level**	**PLI**	**Level**	**ER**	**Level**
3 < Igeo ≤ 4	Heavily contaminated			ER ≥ 320	Very high potential ecological risk
4 < Igeo ≤ 5	Heavily to extremely contaminated				
Igeo > 5	Extremely contaminated				

Table 1.
Classification criteria for different indices.

$$Pollution\ Load\ Index = \left(\frac{C_{Soil}}{C_b}\right) \tag{2}$$

where C^{Se}_{Soil} is the concentration of selenium in any sample x, and C^{i}_{b} is the background concentration of selenium in soils before accumulation which was calculated for each soil by determining the Se concentration in the deepest horizon. The criteria for classifying the pollution load index is presented in **Table 1**.

The potential ecological risk of selenium accumulation to the ecosystem was calculated by the ecological risk index, which was first suggested by Hakanson [33]. The potential ecological risk index was by the following equation:

$$E^i_r = T^i_r \times \left(\frac{C^{Se}_{Soil}}{C^{Se}_b}\right) \tag{3}$$

where T^i_r is the toxic effect of selenium (Se = 10), C^{Se}_{Soil} the concentration of selenium in soil samples, C^{Se}_b is the background concentration of selenium in soils. Classifying criteria is presented in **Table 1**.

6. Health risk assessment

The USEPA approach, which has been extensively used around the world, can be used to assess the health risks associated with heavy metal exposure through food consumption [34]. By calculating the target hazard quotient (HQ) and the hazard index (HI) for selenium, the health risks of ingesting Se will be measured. Below are the equations as follow:

$$THQ = \frac{EF * ED * C_{veg} * IR_{veg}}{BW * AT * RfD} \tag{4}$$

The recommended daily intake (RfD) is the amount of selenium consumed each day through plant-based foods that are deemed to be safe over the course of a lifetime. Depending on the age, sex, and standard tolerable daily intake of Se, the range is 0.02 to 0.075 mg kg1 day1 [35]. EF stands for exposure frequency (365 days per year), ED for exposure duration (74.68 years), C for food's selenium content, IR for food's ingestion rate, BW for average body weight, and AT for an average duration of non-carcinogen exposure (365 days divided by 74.68 years).

The following equation was used to calculate the hazard index (HI) of consuming food while simultaneously absorbing multiple heavy metals:

$$HI = \sum_{i}^{n} THQ_i \tag{5}$$

A negative effect is anticipated to be seen by the exposed population when the HQ/HI values are equal to or higher than 1 [36].

7. Conclusion

Selenium essentiality and toxicity and the narrow range between them made it very critical to keep an eye on selenium deficiency and toxicity through the food chain in humans and animals. While most of the selenium in our food is supplied through soils in our food. Whereas in soils selenium contents depend on soil parent material inheriting different selenium contents through different minerals in the soils. Besides the total selenium contents in soils, other factors also play important role in its availability to plants including, pH, Eh, clay, organic matter, selenium fractions, species, and competing ions which ultimately play a role in its deficiency and toxicity. It is necessary to monitor the food grown in different soils for selenium deficiency or toxicity. While evaluating the degree of toxicity it is necessary to calculate the selenium environmental or ecological risks and health risks associated with high selenium. In the end, it is necessary to consider the soil properties and other factors which influence selenium availability.

Acknowledgements

The authors acknowledge (Prof. Dr. Mohammad Saleem Akhtar, Arid Agriculture University Rawalpindi) for guiding and contributing throughout the research.

Conflict of interest

The authors declare no conflict of interest.

Author details

Muhammad Imran[1*], Zhikun Chen[1], Ayaz Mehmood[2], Shah Rukh[3], Wang Weixie[1], Waleed Asghar[4] and Farhan Iftikhar[5]

1 Key Laboratory of Soil Resource and Biotech Application, Shaanxi Academy of Sciences, Xi'an Botanical Garden of Shaanxi (Institute of Botany Shaanxi Province), Xi'an, China

2 Department of Agricultural Sciences, University of Haripur, Pakistan

3 National Centre of Excellence in Geology, University of Peshawar, Pakistan

4 Faculty of Life and Environmental Sciences, Department of Environmental Sciences, University of Yamanashi, Kofu, Japan

5 School of Soil and Water Conservation, Beijing Forestry University, Beijing, P.R. China

*Address all correspondence to: changwani_baloch2005@yahoo.com

References

[1] Rayman MP. Selenium and human health. Nutrition Reviews. 2012;**379**: 1256-1268

[2] Steinbrenner H, Speckmann B, Pinto A, Sies H. High selenium intake and increased diabetes risk: Experimental evidence for interplay between selenium and carbohydrate metabolism. Journal of Clinical Biochemistry & Nutrition. 2010;**48**(1): 40-45

[3] Yoon JC, Pulgserver P, Chen G, Donovan J, Wu Z, Rhee J, et al. Control of hepatic gluconeogenesis through the transcriptional coactivator PGC-1. Nature. 2001;**413**(6852):131-138

[4] Fordyce F. Selenium deficiency and toxicity in the environment. In: Selinus O, editor. Essentials of Medical Geology. Dordrecht: Springer; 2013. pp. 375-416

[5] N'eve J. Selenium as a 'nutraceutical': How to conciliate physiological and supra-nutritional effects for an essential trace element. Current Opinion in Clinical Nutrition and Metabolic Care. 2002;**5**(6):659-663

[6] Weekley CM, Harris HH. Which form is that? The importance of selenium speciation and metabolism in the prevention and treatment of disease. Chemical Society Reviews. 2013;**42**(23): 8870-8894

[7] Mayland HF, James LF, Panter KE, Sonderegger JL. Selenium in seleniferous environments. In: Jacobs LW, editor. Selenium in Agriculture and Environments. Madison: Soil Science Society of America; 1989. pp. 15-50

[8] Imran M, Akhtar MS, Khan KS, Khalid A, Mehmood A, Rukh S, et al. Total and extractable soil selenium contents variation within and across the parent materials. Journal of Biodiversity and Environmental Sciences. 2016;**9**: 175-186

[9] Tan J. The Atlas of Endemic Diseases and their Environments in People's Republic of China. Marrickville: Science Press; 1989. p. 88

[10] Terry N, Zayed AM, deSouza MP. Tarun AS selenium in higher plants. Annual Review of Plant Physiology and Plant Molecular Biology. 2000;**51**: 401-432

[11] Joy EJM, Broadley MR, Young SD, Black CR, Chilimba ADC, Ander EL, et al. Soil type influences crop mineral composition in Malawi. Science of The Total Environment. 2015;**505**:587-595

[12] Dinh QT, Cui Z, Huang J, Tran TAT, Wang D, Yang W, et al. Selenium distribution in the Chinese environment and its relationship with human health: A review. Environment International. 2018;**112**:294-309

[13] Ramesh R, Subramanian V, Van-Grieken R. Heavy metal distribution in sediments of Krishna river basin, India. Environmental Geology and Water Sciences. 1990;**15**:207-216

[14] Merrill DT, Manzione MA, Peterson IJ, Parker DS, Chow W, Hobbs AO. Field evaluation of arsenic and selenium removal by iron coprecipitation. Journal - Water Pollution Control Federation. 1986;**58**: 18-26

[15] Balistrieri LS, Chao TT. Selenium adsorption by goethite. Soil Science Society of America Journal. 1987;**51**: 1145-1151

[16] Balistrieri LS, Chao TT. Adsorption of Se by amorphous iron oxyhydroxide and manganese dioxide. Geochimica et Cosmochimica Acta. 1990;**54**(3):739-751

[17] Sarquis M, Mickey CD. Selenium. Part 1: Its chemistry and occurrence. Journal of Chemical Education. 1980;**57**: 886-889

[18] Gissel-Nielsen G, Bisbjerg B. The uptake of applied selenium from soils by plants. 2. The utilization of various selenium compounds. Plant and Soil. 1970;**32**:382-396

[19] Eisler R. Selenium hazards to fish, wildlife, and invertebrates: A synoptic review. U.S. Fish and Wildlife Servo Biol. Rep. 1985; 85(1.5). U.S. Gov. Print. Office, Washington, DC

[20] Vokal-Borek H. Selenium, Univ. of Stockholm Inst. of Physics Rep. Sweden: Stockholm; 1979. pp. 79-16

[21] Ryden IC, Syers JK, Tillman RW. Inorganic anion sorption and interactions with phosphate sorption by hydrous ferric oxide gel. Journal of Soil Science and Plant Nutrition;**187**(38):211-217

[22] Weast RC. Handbook of Chemistry and Physics. 66th ed. Ceveland, OH: The Chemical Rubber Co.; 1989

[23] Lakin HW. Selenium accumulation in soils and its absorption by plants and animals. Geological Society of America Bulletin. 1972;**83**:181-189

[24] Elrashidi MA, Adriano DC, Workman SM, Lindsay WL. Chemical equilibria of selenium in soils: A theoretical development. Soil Science. 1987;**144**:141-152

[25] Cutter GA. Selenium in reducing waters. Science (Washington, DC). 1982; **217**:829-831

[26] Cutter GA, Bruland KW. The marine biogeochemistry of selenium: Are-evaluation. Limnology and Oceanography. 1984;**29**:1179-1192

[27] Sarathchandra SU, Watkinson IH. Oxidation of elemental selenium to selenite by Bacillus megaterium. Science (Washington, DC). 1981;**21**:600-601

[28] Rosenfeld I, Beath OA. Selenium: Geobotany, Biochemistry, Toxicity, and Nutrition. New York: Academic Press; 1964

[29] Tolu J, LeHécho I, Bueno M, Thiry Y, Potin-Gautier M. Selenium speciation analysis at trace level in soils. Analytical Chemistry Acta. 2011;**684**:126-133

[30] Imran M, Akhtar MS, Mehmood A, Rukh S, Khan A, Zhikun C, et al. Soil selenium transformation across different parent materials in Pothwar uplands of Pakistan. Arabian Journal of Geosciences. 2020;**13**:1098. DOI: 10.1007/s12517-020-06111-1

[31] Muller G. Schwermetalle in den sediments des RheinsVeranderungen seitt 1971. Umschan. 1979;**79**:778-783

[32] Zhang P, Qin C, Hong X, Kang G, Qin M, Yang D, et al. Risk assessment and source analysis of soil heavy metal pollution from lower reaches of Yellow River irrigation in China. Science of The Total Environment. 2018;**633**:1136-1147

[33] Hakanson L. An ecological risk index for aquatic pollution control. Sedimentol Approaches Water Research. 1980;**14**: 975-1001

[34] Xiao R, Guo D, Ali A, Mi S, Liu T, Ren C, et al. Accumulation, ecological-health risks assessment, and source apportionment of heavy metals in paddy soils: A case study in Hanzhong, Shaanxi,

China. Environmental Pollution. 2019; **248**:349-357

[35] USEPA (United States Environ. Prot. Agency). Integrated Risk Information System. 2015. Available from: http://www.epa.gov/iris

[36] Cao C, Chen XP, Ma ZB, Jia HH, Wang JJ. Greenhouse cultivation mitigates metal-ingestion-associated health risks from vegetables in wastewater-irrigated agroecosystems. Science of the Total Environment. 2016; **560**:204-211